Energy Inequalities: Crises, Pricing, and Policies

RIVER PUBLISHERS SERIES IN ENERGY SUSTAINABILITY AND EFFICIENCY

Series Editor

PEDRAM ASEF
Lecturer (Asst. Prof.) in Automotive Engineering,
University of Hertfordshire,
UK

The "River Publishers Series in Sustainability and Efficiency" is a series of comprehensive academic and professional books which focus on theory and applications in sustainable and efficient energy solutions. The books serve as a multi-disciplinary resource linking sustainable energy and society, fulfilling the rapidly growing worldwide interest in energy solutions. All fields of possible sustainable energy solutions and applications are addressed, not only from a technical point of view, but also from economic, social, political, and financial aspects. Books published in the series include research monographs, edited volumes, handbooks and textbooks. They provide professionals, researchers, educators, and advanced students in the field with an invaluable insight into the latest research and developments.

Topics covered in the series include, but are not limited to:

- Sustainable energy development and management;
- Alternate and renewable energies;
- Energy conservation;
- Energy efficiency;
- Carbon reduction;
- Environment.

For a list of other books in this series, visit www.riverpublishers.com

Energy Inequalities: Crises, Pricing, and Policies

George Xydis
University of the Peloponnese,
Patras, Greece

Aarhus University,
Herning, Denmark

River Publishers

Routledge
Taylor & Francis Group
NEW YORK AND LONDON

Published 2026 by River Publishers
River Publishers
Broagervej 10, 9260 Gistrup, Denmark
www.riverpublishers.com

Distributed exclusively by Routledge
605 Third Avenue, New York, NY 10158, USA
4 Park Square, Milton Park, Abingdon, Oxon OX14 4RN

Energy Inequalities: Crises, Pricing, and Policies / George Xydis.

Routledge is an imprint of the Taylor & Francis Group, an informa business

ISBN 978-87-4380-890-9 (hardback)
ISBN 978-87-4380-895-4 (paperback)
ISBN 978-87-4380-893-0 (online)
ISBN 978-8-743-80892-3 (ebook master)

DOI: 10.1201/9788743808923

Contents

Preface

The global energy landscape is marked by huge inequalities, where access, affordability, and resilience are increasingly divided. The book *Energy Inequalities: Crises, Pricing, and Policies* confronts these disparities, offering a rigorous examination of how energy crises, market structures, and policy decisions impact the most vulnerable. Drawing on case studies from European households, grid operators, and emerging energy communities, this book bridges empirical research with actionable policy insights to chart a more equitable energy future.

At the heart of this volume is the 2022 energy crisis, which forced Czech households to adopt urgent energy-saving measures – often at great personal sacrifice. Yet, as Chapter 2 reveals, these adaptations were far from uniform, exposing deep divides in household flexibility and financial security. Our society has not yet worked in a structured way to confront these disparities. Another case presented focuses on the rapid rise of electric vehicles, explored in Chapters 3 and 4, introducing new challenges: while EVs promise cleaner transportation, their integration risks reinforcing existing inequities, particularly for those reliant on public charging or living in grid-constrained areas.

Energy communities, such as those emerging in Poland (Chapter 5), represent a potential counterbalance – a grassroots approach to democratizing energy access. But their success hinges on overcoming regulatory inertia and entrenched market dynamics. To ground these discussions in practical application, the book concludes with targeted exercises (Appendix A) that challenge students (and policymakers) to – first off – develop the mindset and to quantify disparities and model inclusive solutions.

This book is a call to action. By dissecting the root causes of energy inequality – from volatile pricing to exclusionary infrastructure – we aim to equip readers with the tools to build systems that prioritize justice alongside sustainability. The path forward demands more than technical innovation; it requires a fundamental rethinking of who benefits from energy transitions, and who bears their costs.

George Xydis
June 2025, Nafpaktos, Greece

List of Contributors

Krolikowska, Aleksandra, *Department of Business Development and Technology, Aarhus University, Denmark*

Panasik, Milosz, *Department of Business Development and Technology, Aarhus University, Denmark*

Paramasivan, Prabakar, *Department of Business Development and Technology, Aarhus University, Denmark*

Vanda Kuti, Vivien, *Department of Business Development and Technology, Aarhus University, Denmark*

Volny, Jakub, *Department of Business Development and Technology, Aarhus University, Denmark*

Xydis, George, *Department of Mechanical Engineering, University of the Peloponnese, Greece; Department of Business Development and Technology, Aarhus University, Denmark*

List of Figures

List of Tables

List of Abbreviations

EEC	Eisall Energy Cooperative
EC	Energy communities
ICT	Information and communication technologies
IEA	International Energy Agency
PHEV	Plug-in hybrid electric vehicles
RES	Renewable energy sources
RTP	Real-time pricing
TSO	Transmission system operator

1

Introduction

George Xydis[1,2]

[1]Department of Mechanical Engineering, University of the Peloponnese, Greece
[2]Department of Business Development and Technology, Aarhus University, Denmark

1.1 Energy Inequalities

Energy inequalities represent one of the most pressing challenges in the global transition to sustainable energy systems. These disparities manifest in unequal access to affordable energy, significant exposure to price volatility, and exclusion from the benefits of clean energy technologies (Volodzkiene & Streimikiene, 2023). Vulnerable households, marginalized communities, and developing regions often bear the weight of energy crises, while wealthier populations and corporations receive the rewards of energy transitions (Acheampong et al., 2021). This book examines the root causes of these inequities – from market structures and policy failures to infrastructural barriers – and explores how they can be addressed through inclusive governance, fair pricing mechanisms, and community-driven solutions.

1.2 Book Motivation

The motivation behind the book *Energy Inequalities: Crises, Pricing, and Policies* stems from the growing divide in energy access and affordability, exacerbated by recent crises such as the 2022 energy shock (Ruhnau et al., 2023; Xydis & Nanaki, 2023). While much of the discourse on energy transitions focuses on technological innovation and decarbonization, far less attention is paid to who benefits and who is left behind (Sovacool et al., 2019). This book seeks to fill that gap by providing empirical evidence, real-world case studies, and actionable policy insights. By highlighting the lived

DOI: 10.1201/9788743808923-1

experiences of households, the challenges of integrating electric vehicles, and the potential of energy communities (Niswander & Xydis, 2024), the book underscores the urgent need for a justice-centered approach to energy policy.

1.3 Objectives and Aim of the Book

The primary objective of *Energy Inequalities: Crises, Pricing, and Policies* is to analyze the systemic drivers of energy inequality and propose pathways toward a more equitable energy future. It aims to:

- Investigate how energy crises disproportionately affect vulnerable populations,
- Assess the socio-economic barriers to adopting clean energy technologies,
- Evaluate the role of policy and market design in perpetuating or mitigating disparities, and
- Highlight grassroots solutions, such as energy communities, that empower marginalized groups.

By bridging academic research with practical policy recommendations, the book strives to equip policymakers, researchers, and advocates with the tools needed to design inclusive energy systems.

1.4 Book Structure

The book is structured to first diagnose the problem of energy inequality before exploring solutions. It opens with an in-depth analysis of household responses to the 2022 energy crisis, revealing stark disparities in resilience and adaptability. Subsequent chapters examine the challenges of electric vehicle (EV) integration, where pricing models and infrastructure gaps risk reinforcing existing inequities. The discussion then shifts to the transformative potential of energy communities, particularly in Poland, where regulatory and financial hurdles slow progress. Finally, the book concludes with practical exercises designed to help readers quantify energy disparities and model equitable policy interventions. Through this structured approach, the book moves from critique to action, offering a roadmap for a just energy transition.

References

Acheampong, A. O., Dzator, J., & Shahbaz, M. (2021). Empowering the powerless: does access to energy improve income inequality?. Energy Economics, 99, 105288.

Niswander, G., & Xydis, G. (2024). California's electric grid nexus with the environment. International Journal of Emerging Electric Power Systems, 25(6), 723-735.

Ruhnau, O., Stiewe, C., Muessel, J., & Hirth, L. (2023). Natural gas savings in Germany during the 2022 energy crisis. Nature Energy, 8(6), 621-628.

Sovacool, B. K., Martiskainen, M., Hook, A., & Baker, L. (2019). Decarbonization and its discontents: a critical energy justice perspective on four low-carbon transitions. Climatic Change, 155, 581-619.

Volodzkiene, L., & Streimikiene, D. (2023). Energy inequality indicators: a comprehensive review for exploring ways to reduce inequality. Energies, 16(16), 6075.

Xydis, G., & Nanaki, E. (2023). Energy Industry Transition Transformation in the Wake of COVID-19. Energies, 16(21), 7455.

2

Energy-Saving Behavior in Czech Households During the Energy Crisis in 2022

Jakub Volny[1]

[1]Department of Business Development and Technology, Aarhus University, Denmark

Abstract

In early 2022, the Russian invasion of Ukraine triggered a significant rise in global energy prices, impacting households across Europe, including the Czech Republic. This study examines how these circumstances influenced the adoption of low-tech and no-tech energy-saving behaviors among Czech households. Using data from 116 respondents, the research identifies common practices and the demographic and socio-economic factors driving these behaviors. Findings reveal that older individuals, women, and lower-income households were more likely to adopt energy-saving measures. The study also notes increased awareness and interest in energy-saving practices, indicating potential for long-term shifts in energy consumption. However, fewer households invested in substantial measures like retrofitting, highlighting the need for policies supporting both immediate and long-term energy efficiency. This research provides valuable insights for policymakers and community leaders to enhance household energy efficiency.

2.1 Introduction

In early 2022, the Russian invasion of Ukraine triggered a substantial escalation in global energy prices, significantly impacting households across Europe with increased energy costs and introducing sharp economic constraints. This energy crisis has brought the importance of energy-saving behaviors into the spotlight, particularly in the context of household energy

DOI: 10.1201/9788743808923-2

management (Hoffmann et al., 2023). Czech households, like many others, faced the challenge of maintaining their energy needs while adjusting to rapidly increasing prices. This paper seeks to investigate how these geopolitical and economic circumstances influenced the adoption of low-tech and no-tech energy-saving behaviors among Czech households.

Understanding energy-saving behavior is crucial not only for managing short-term crises but also for fostering long-term strategies to enhance energy resilience and sustainability since residential consumption represents around one-fifth of the total energy consumption (Brounen et al., 2013). Low-tech and non-tech solutions, which include measures that do not require significant technological input, such as adjusting thermostat settings, utilizing natural light, or modifying cooking and washing habits, represent a practical and immediately implementable way of reducing energy consumption (Alexander & Yacoumis, 2018). These behaviors can be pivotal in mitigating the impact of sudden energy price increases on household budgets and in contributing to broader energy conservation efforts.

This paper has two focus points. Firstly, it aims to investigate the range of energy-saving behaviors adopted by Czech households in response to the 2022 energy crisis. Secondly, it seeks to identify the demographic and socio-economic factors influencing these behaviors, providing insights into how different segments of the population respond to energy crises. Through an online survey distributed across various regions of the Czech Republic, this study will explore the correlation between the energy crisis and the adoption rate of energy-saving practices.

By analyzing the shifts in household energy management during a critical period, this research contributes to the broader discourse on sustainable living practices and energy policy. The findings aim to inform policymakers, energy providers, and community leaders, providing evidence-based recommendations to support households in enhancing their energy efficiency under normal and crisis conditions.

In practice, in this work, the problem statement could be "How did the 2022 energy crisis, triggered by the Russian invasion of Ukraine, influence the adoption of low-tech and no-tech energy-saving behaviors among Czech households, and what demographic and socio-economic factors affected these behaviors?"

2.2 Theoretical Background

2.2.1 2022 Energy crisis in Europe

In 2022, Europe faced a significant energy crisis that had profound impacts on energy prices, supply security, and prompted shifts in energy policy across European nations. The crisis can be mostly attributed to geopolitical tensions, specifically the Russia-Ukraine war, combined with existing pressures from the COVID-19 pandemic – even in other sectors (Koumoulides et al., 2022). As countries rebounded from the pandemic, there was a sudden increase in energy demand, which was not met by an equivalent supply, leading to increased energy prices (Ozili & Ozen, 2023). The invasion of Ukraine by Russia in early 2022 worsened the situation, as Russia was a major supplier of natural gas and oil to Europe. Sanctions imposed on Russia by the EU and other Western entities disrupted these supplies further, dramatically impacting the energy market (Jing, 2023). Europe's heavy reliance on Russian energy imports made it particularly vulnerable to these disruptions. The conflict highlighted the fragility of European energy security and the risks associated with geopolitical dependencies (Zhou et al., 2023).

The crisis resulted in energy prices increasing, with some reports indicating a rise of 62.6–112.9% in 2022 (Guan et al., 2023). This price surge was felt most acutely in natural gas and electricity markets. The increased cost of energy contributed to inflationary pressures across the continent, straining economies still recovering from the pandemic effects. Industries heavily dependent on energy, such as manufacturing and utilities, were particularly hit. Higher energy costs led to increased living costs, pushing more individuals into energy poverty. This situation has had wider social implications, including increased dissatisfaction with political handling of the crisis and stronger support from populist political parties. (Havlík & Kluknavská, 2023)

2.2.2 Energy-saving behavior

As the energy bills got higher, energy-saving behavior naturally occurred in European households. In the Czech Republic specifically, electricity consumption of households decreased on average by 9% from the previous year (the biggest decrease of 17.9% in June, and 16.7% in May) (Energetický regulační úřad, 2023a), and natural gas consumption of households decreased by 20% (Energetický regulační úřad, 2023b). To define the energy-saving behavior, it is "the daily and habitual practices of households that focus on specific reductions in energy use" (Trotta, 2018). This research paper focuses on low-tech and no-tech energy-saving practices. As Alexander & Yacoumis (2018) propose, these practices are rooted in a 'simple living' approach, which focuses on saving energy through behavioral change instead of relying on advanced technology that, while potentially reducing energy consumption, causes growth in material demand and energy dependence. Handwashing dishes instead of using a dishwasher, line drying clothes instead of using a dryer, or reducing the heating temperature in winter are some of the examples of low-tech/no-tech energy-saving behaviors. Energy-efficiency consideration, such as investing in heat insulation of a house or replacing lightbulbs with more efficient ones, is included in the low-tech and no-tech category in thischapter, though it requires additional expenditure.

The energy-saving behavior of individuals is highly subjective and there is a number of scientific papers investigating the factors like demographic profile or environmental beliefs that influence it. For example, Hori et al. (2013) investigated the influence of global warming consciousness, environmental behavior, social interaction, age, and income on the individuals' energy-saving behavior in Asia. According to their study, age and income did not have any significant influence, while the three other factors had a positive correlation with energy-saving behavior. Trotta (2018) looked into factors such as age, marital status, education level, dwelling type, and pro-environmental attitude and knowledge among Brits. He observed that even though the income does not correlate with energy-saving behavior, households with lower income tend to be more flexible with energy demand, meaning that they consume less energy if the prices are high by adjusting their behavior. Similarly, as in Hori et al. (2013), the pro-environmental behavior, attitude, and knowledge positively correlated with energy-saving behaviors. Moreover, married respondents were associated with higher adoption of energy-saving behavior (Trotta, 2018). Education level was not proven as a predictor of energy-saving behaviors (Sardianou, 2007; Trotta, 2018).

The positive correlation between marital/family status and energy-saving behavior was also confirmed by Belaïd & Joumni (2020), who investigated the influential factors among French citizens and found that families with children are more likely to engage in energy-saving. In their research, a political or ecological ideology does not seem to have influence; however, being part of a civil society organization, a union, or a political party was observed to have a positive effect on energy-saving attitudes. (Belaïd & Joumni, 2020) The role of gender in energy-saving behavior is rather conflicted, as some studies neglect its influence (Martinsson, 2011; Trotta, 2018) while other studies show that women are more likely to engage in energy-saving. (Räty & Carlsson-Kanyama, 2010; Belaïd & Joumni, 2020) Impact of ownership, on the other hand, is clearly positive, showing that people who own their housing are more likely to adopt energy-saving behavior (Belaïd & Joumni, 2020; Trotta, 2018).

2.2.3 Energy-saving behavior in the energy crisis

Scientific literature investigating the impact of energy crisis on energy-saving behavior has continuously emerged since 2022. As pointed out by Tóth et al. (2023), who looked into the energy consumption of Hungarian households during the energy crisis, low-income households exhibit a smaller change in energy-saving behavior in the crisis scenario. That is because they often fall into energy poverty, and they already strictly limit their energy use, so there is no further space for savings. Middle-income households seem to be the most responsive to the increase in energy prices. Moreover, Tóth et al. (2023) notice that high gas-consumption areas tended to be clustered together, indicating that geographical and community characteristics influence energy use patterns as well. The community characteristics involve energy literacy and awareness, which prove to have a positive correlation with energy-saving (Tóth et al., 2023).

Liobikien et al. (2023) investigated how the crises such as COVID-19 pandemic, the war in Ukraine, and climate change impact the energy-behavior of households in Lithuania. It was identified that only half of the respondents agreed that their lower individual consumption as a whole can contribute and positively affect climate change. This recognition of the opportunity to make a change as an individual is mentioned as one of the influential factors for energy-saving behavior. Furthermore, Liobikien et al. (2023) point out that the pandemic increased concern about the environment among respondents. It is also discussed that even though respondents say the pandemic made them save energy more than the war, the energy consumption data indicate the opposite, but that might be caused by the fact that people spent considerably more time at home during the pandemic. Nevertheless, their research shows that even though the war in Ukraine and the COVID-19 pandemic are not directly related to environmental issues, they have a significant impact on their perception.

Kola-Bezka & Leki (2024) have focused on the specific energy-saving practices and influence of the socio-economic and demographic characteristics of Polish households. Contrary to some of the previously mentioned studies, Kola-Bezka & Leki (2024) propose that the higher the age of a respondent, the higher the adoption rate of energy-saving behavior. Moreover, it indicates that low-income households are more likely to engage in practices related to behavir change while high-income households are more likely to implement some of the cost-intensive measures, such as heat insulation retrofitting. Another significant factor was the energy carrier used for heating (Tin & Xydis, 2025). Households that used natural gas or hard coal adopted energy-saving behavior more. As for the specific practices, the most common ones were turning off the appliances and lights when not used, reducing the frequency of clothes washing, or lowering the temperature when no one is home. (Kola-Bezka & Leki, 2024).

2.3 Methodology

The research is based on primary data from 116 respondents, collected through an online survey distributed mainly on social media. The first part of the survey is concerned with socioeconomic and demographic information about the respondent, such as gender, age, education level, or field of employment. The subsequent second chapter then collects information about the respondent's household, such as the dwelling type, number of people living in the household, which home appliances are present in the household, or the extent to which the increased energy prices impacted the household's finances.

The third part investigates the individual energy-saving practices implemented in households after the energy crisis started. The selection of specific practices was inspired by research papers that investigated similar topics in the past to allow potential comparison. (Hall et al., 2013; Hori et al. 2013; Belaïd & Joumni, 2020; Kola-Bezka & Leki, 2024;) Similarly, as in the research paper by Hall et al. (2013), who clustered the practices based on the physical area in the household, the practices are grouped into four categories: heating, personal hygiene, dishwashing and laundry, and other home appliances. Lastly, the fourth part of the survey looks into respondents' personal attitudes and perceptions towards energy-saving, such as perceived change in energy consumption awareness, perceived potential for further energy-saving, or recognition of the positive impact of the household's energy-saving on the environment. The overview of the information to be collected can be seen in Table 2.1. The survey was written in Czech in order to avoid a language barrier since not all respondents would understand English. A translated version of the survey can be found in Appendix 1.

Firstly, the data will be analysed to identify the most adopted energy-saving practices among the Czech households during the energy crisis. Moreover, the impact of the increased energy prices on Czech households' finances as well as their personal attitudes and perceptions towards energy saving will be investigated in the context of the energy crisis. Lastly, the influence of individual independent variables (demographic characteristics, household information, personal attitudes, and perceptions) on the dependent variables (energy-saving practices) will be investigated. The average number of adopted practices relative to each of the independent variables will be

Table 2.1 Survey structure.

Category	Collected information
Demographic characteristics	• Age • Gender • Field of employment • Education level • Income level
Household information	• Dwelling type • Ownership • Number of inhabitants • Person responsible for energy bill payment • Energy carrier used for heating • Energy carrier for water heating • The extent to which the household finances were affected by increased electricity prices • The extent to which the household finances were affected by increased heating prices • Appliances present in the household
Energy-saving practices implemented during the energy crisis	• Related to heating • Related to personal hygiene • Related to dishwashing and laundry • Related to other home appliances
Personal attitudes and perceptions towards energy saving	• Perceived change in interest in energy saving • Perceived change in awareness of energy consumption • Continuity of the energy-saving behavior even after the decrease in energy prices • Potential for further energy-saving • Role of financial motivation for energy-saving • Recognition of the positive impact of household energy-saving on the environment

visualized to identify whether there is a correlation. The independent variables with significant influence will be discussed.

2.4 Results

2.4.1 Adoption rate of energy-saving practices

The results regarding the adoption of the specific energy-saving practices since the energy crisis started can be seen in Figures 2.1–2.4. From the

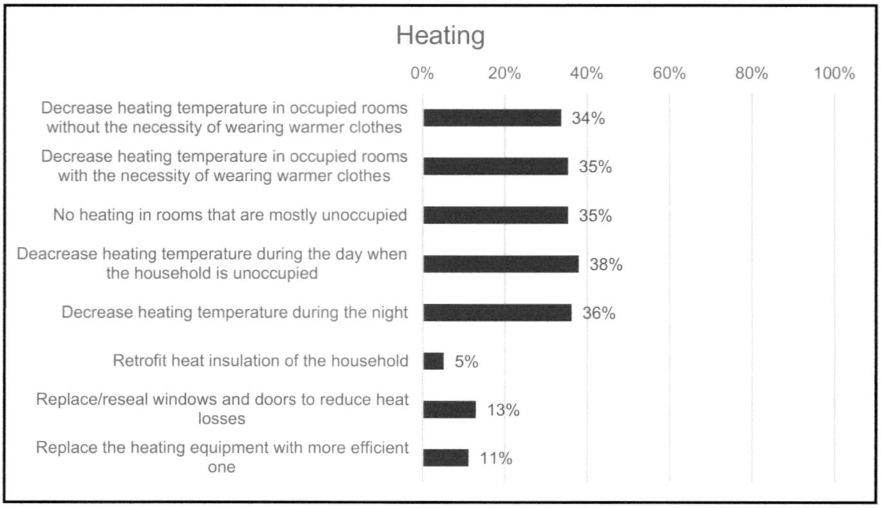

Figure 2.1 Occurrence of energy-saving practices related to heating.

heating category (Figure 2.1), the most utilized practices were related to temperature regulation. Decreasing temperature in the whole household during the night, or during the day when the household is not occupied, ranked the highest. At least 35% of respondents decreased the temperatures to an extent when it disturbed their thermal comfort, so they had to wear warmer clothes inside their household. Behavior related to the replacement or retrofitting of inefficient equipment ranked significantly lower.

In the dishwashing and laundry category (Figure 2.2), most respondents started to be more cautious about doing laundry and washing dishes efficiently, as 41% stated that they started using the laundry machine less often or only when it was fully loaded, and 36% stated the same about the dishwasher. In the case of the usage of a clothes dryer, the percentage is lower; however, that is caused by the lower number of respondents owning a dryer (only 43% own one). Other practices that ranked relatively high are shortening washing time/reducing water temperature when using a laundry machine (18%) and switching to line drying from using an automatic clothes dryer (16%). Replacing the washing machine, laundry machine, or automatic dryer with a more efficient one, as well as switching to doing laundry manually, ranked the lowest.

In the case of personal hygiene (Figure 2.3), most people started reducing their showering time (42%) and switching to showering instead of taking a bath (31%). These two are followed by the installation of tap aerators or

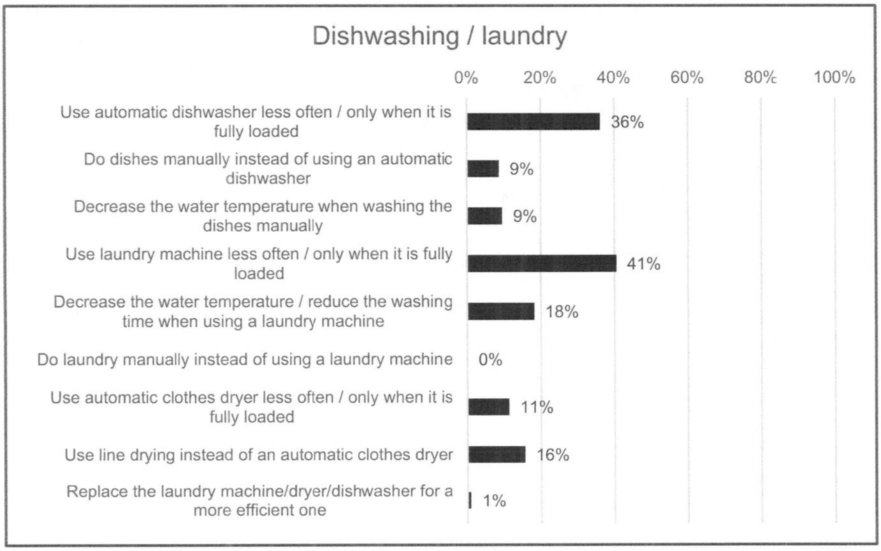

Figure 2.2 Occurrence of energy-saving practices related to personal dishwashing/laundry.

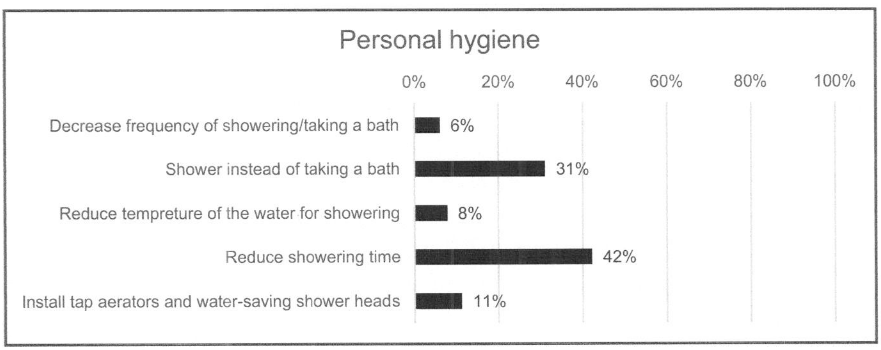

Figure 2.3 Occurrence of energy-saving practices related to personal hygiene.

water-saving shower heads (11%), a decrease in water temperature during showering (8%), and lastly, decreased frequency of showering/taking a bath.

The energy-saving practices in the "other home appliances" category (Figure 2.4) reach the highest adoption rates. Most respondents state that they began turning off the TV and PC if it is not actively used (51% and 42% respectively) and replaced their installed lightbulbs with more efficient ones (46%). Avoiding unnecessary usage of artificial light during daylight ranked closely after (40%). Unplugging home appliances when not being used and adjusting cooking for energy saving was adopted far less, with 21% and 20%

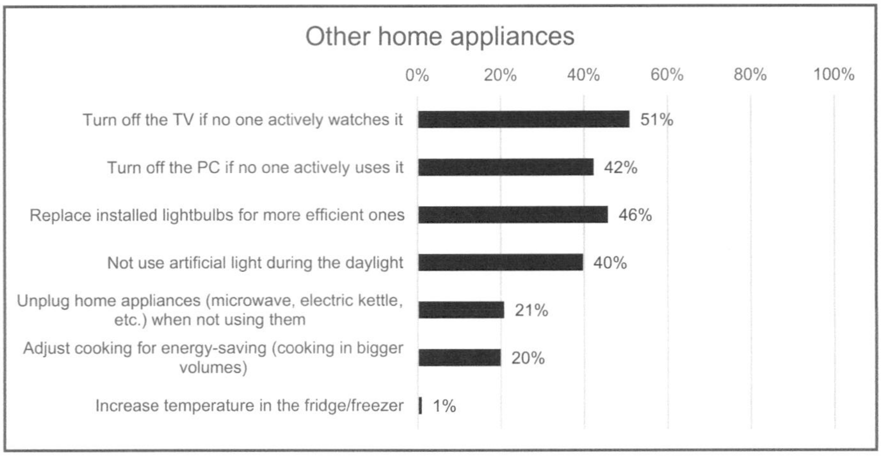

Figure 2.4 Occurrence of energy-saving practices related to other home appliances.

adoption rates. Increased temperature in the fridge or freezer to a still acceptable temperature ranked the lowest.

2.4.2 Impact of the energy crisis on Czech households

To understand the severity of the crisis' financial impact on Czech households, respondents were asked to assess the impact of the increased prices for heating (Figure 2.5) and electricity (Figure 2.6) on a scale of 1 (none) to 5 (critical). In both cases, similar results can be observed. Around 6-7% were critically impacted and 14–17% were significantly impacted. 33–37% of respondents experienced medium impact on their household finances, 36–37% were affected only slightly, and 6–7% were not affected at all.

To get an insight into the respondents' attitudes and beliefs, and how the energy crisis might have changed them, they were asked to express agreement/disagreement with six statements. Answers about the first statement (Figure 2.7) reflect a positive change in interest of the respondents in energy-saving practices, as 21% agreed and 56% agreed that the energy crisis made them more interested in how to save energy in their households. In the case of the second statement (Figure 2.8) concerned with energy consumption awareness, the change is not that significant, as 17% agreed and 36% rather agreed that they became more aware of their energy consumption since the energy crisis started.

The fourth statement (Figure 2.9) investigates potential continuity of the implemented practices after the energy crisis. In this case, responses are very

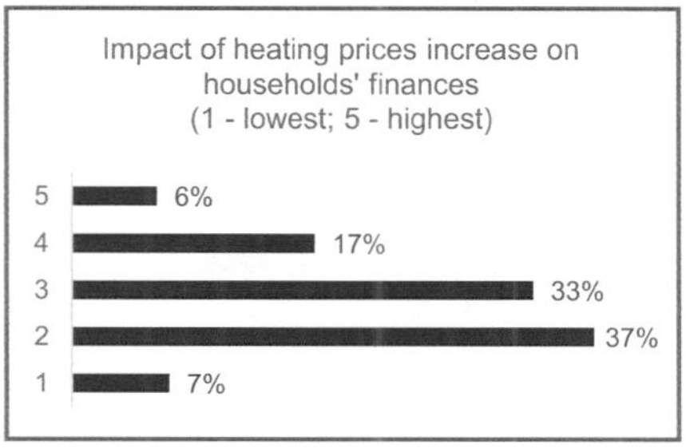

Figure 2.5 Impact of heating prices increase.

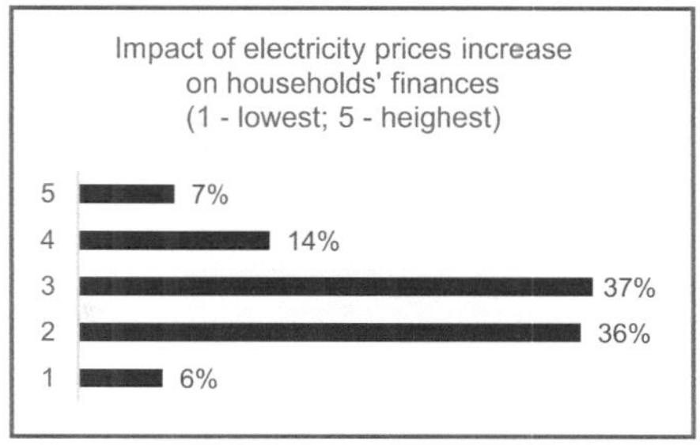

Figure 2.6 Impact of electricity prices increase.

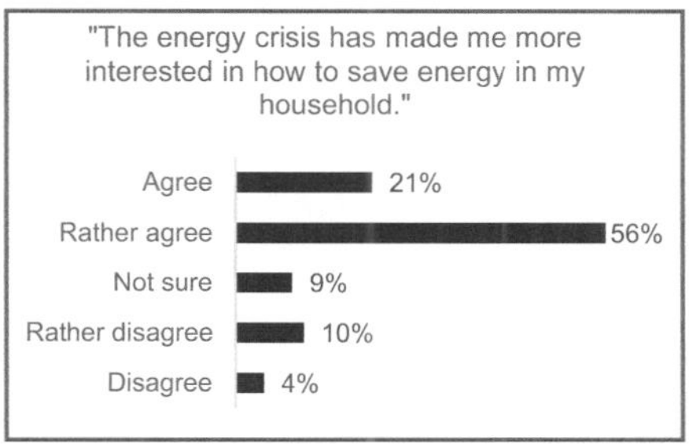

Figure 2.7 Energy-saving interest.

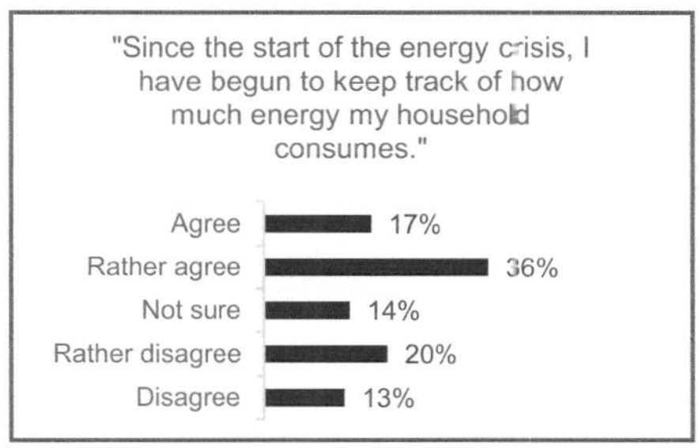

Figure 2.8 Energy consumption awareness.

positive as 39% agree and 41% rather agree that the adopted practices will be maintained even when the energy prices are lower again. Fifth statement (Figure 2.10) explores the potential space for further energy savings without significantly affecting one's comfort. The data collected indicate that 53% of respondents (16% agree; 37% rather agree) could still reduce their energy consumption without having a significant negative impact on their comfort. On the other hand, 14% of respondents would be essentially affected by further energy-saving, representing part of the population at a potential risk of falling into energy poverty.

The fifth statement reflects how important the role of financial motivation is when it comes to energy saving. Around 25% of respondents agree, and 34% rather agree that their motivation for reducing their energy consumption is purely financial. On the other hand, 10% disagree and 22% rather disagree with that statement (Figure 2.11). The sixth statement then looks into whether respondents recognize the environmental importance of energy saving in households. Only 12% of respondents either disagree (4%) or rather disagree (8%) that reducing household energy consumption has a significant

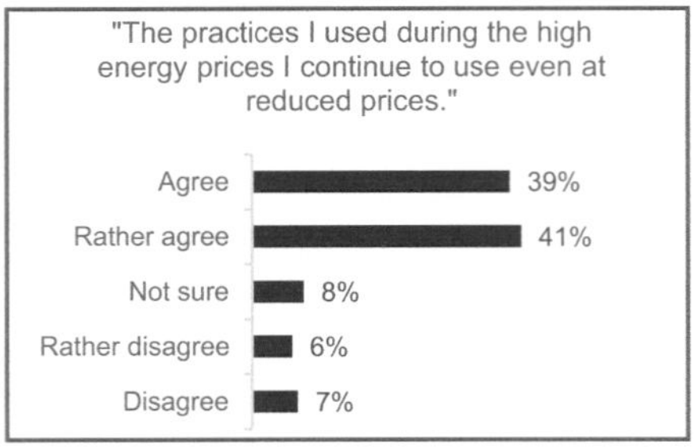

Figure 2.9 Continuity of the practices.

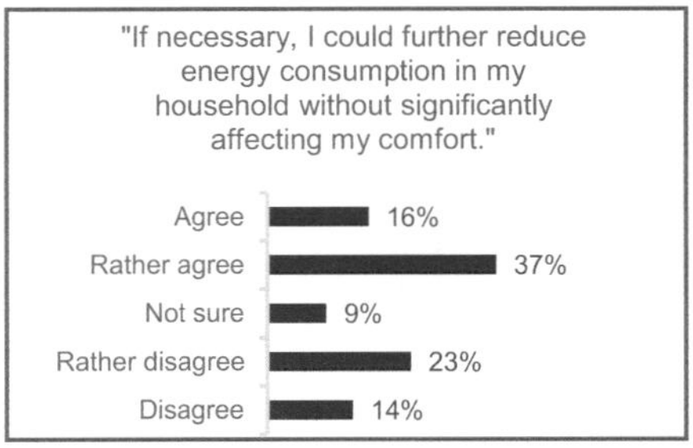

Figure 2.10 Potential for further energy savings.

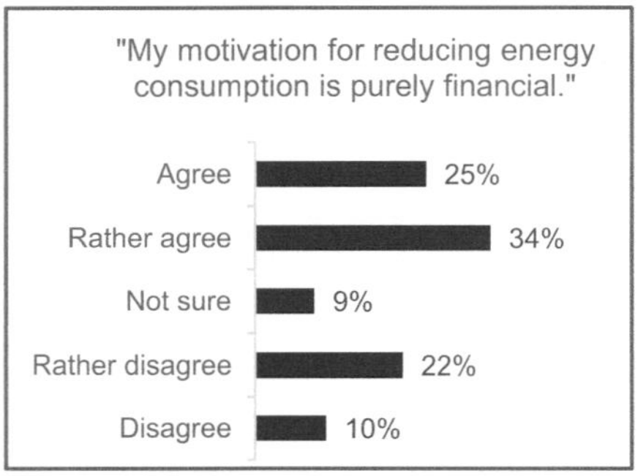

Figure 2.11 Financial motivation.

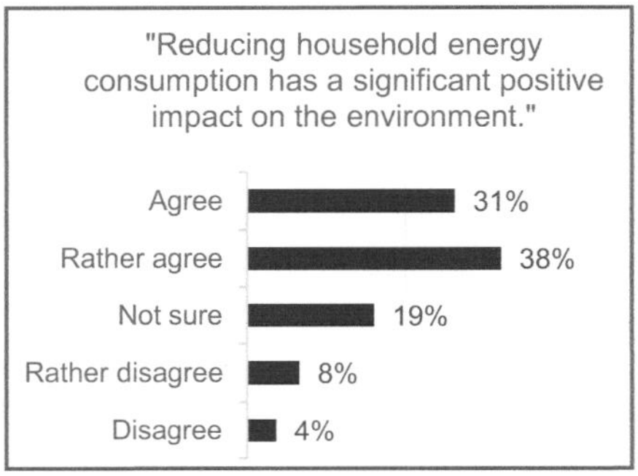

Figure 2.12 Positive impact on the environment.

positive impact on the environment. At least 31% agree and 38% rather agree with the statement, reflecting growing recognition of environmental issues (Figure 2.12).

2.4.3 Influence of demographic characteristics, household details, and personal attitudes on energy-saving behavior

After visualizing the average number of implemented practices per respondents' demographic characteristics, household details, and personal attitudes, few correlations between these independent and dependent variables can be observed. Even though there is no linear dependency between age and average number of adopted practices (Figure 2.13), the data shows that people over 55 years old have the highest adoption rate, followed by the age group of 26–35. In case of biological sex of the respondent (Figure 2.14), females seem to adopt more energy-saving practices than males during the energy crisis.

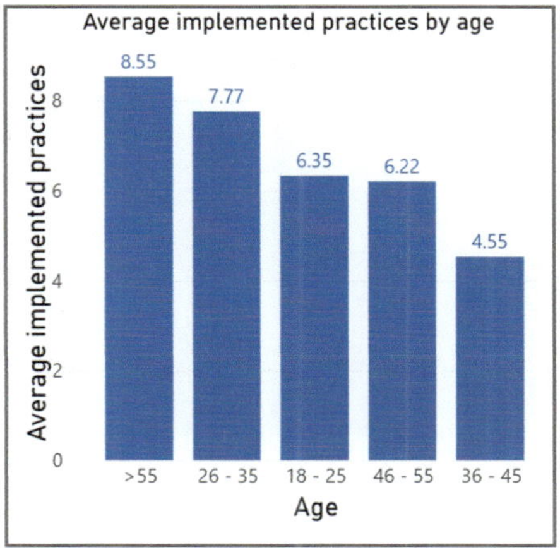

Figure 2.13 Influence of age.

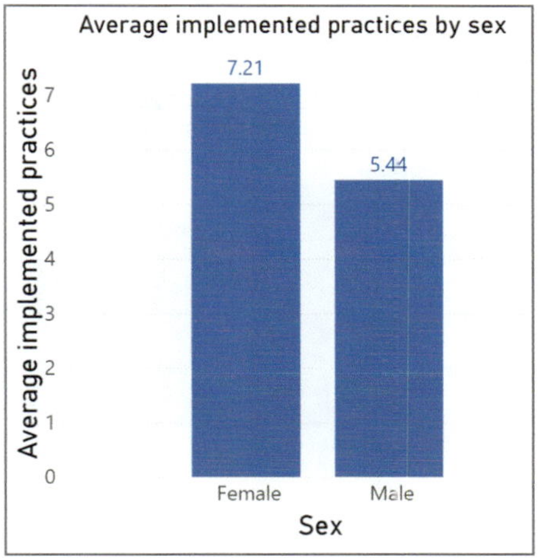

Figure 2.14 Influence of sex.

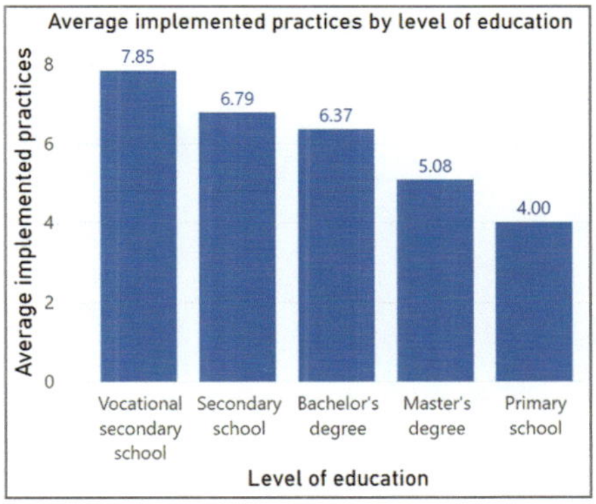

Figure 2.15 Influence of education level.

As for the influence of education level (Figure 2.15), respondents with vocational education report slightly higher adoption rates (7.85), followed by secondary school graduates (6.79), and higher education (bachelor – 6.37; master – 5.08). Respondents with primary education rank the lowest, with only four adopted practices on average.

In case of the dwelling type (Figure 2.16), respondents living in semi-attached, terraced houses, and apartments rank the highest in adoption of the practices, but the differences are rather small. Ownership of the household (Figure 2.17) does not seem to have such a high influence as respondents living in a rented dwelling report a higher adoption rate by only 0.71 in comparison to respondents owning their home.

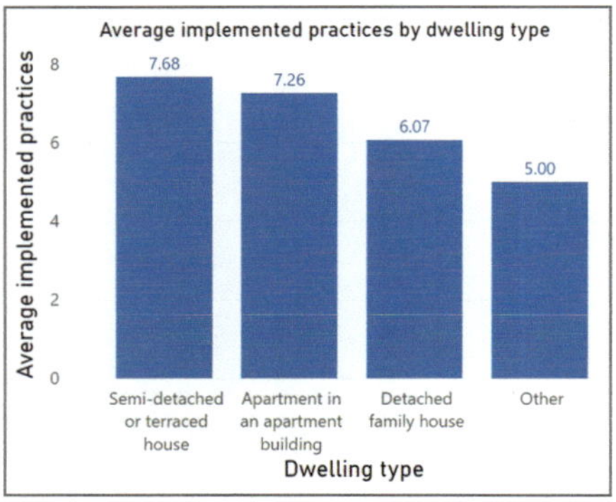

Figure 2.16 Influence of dwelling type.

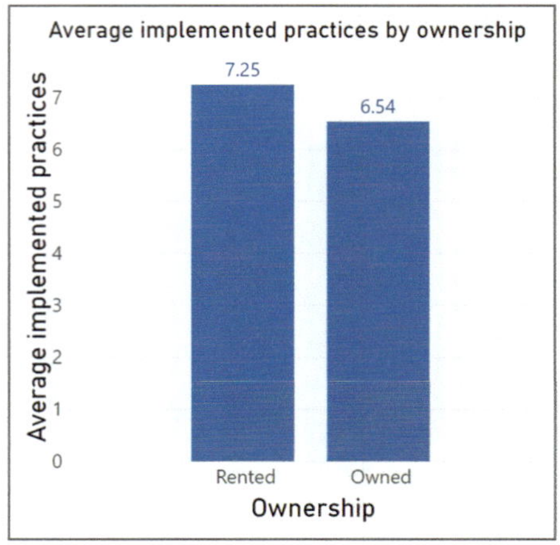

Figure 2.17 Influence of ownership.

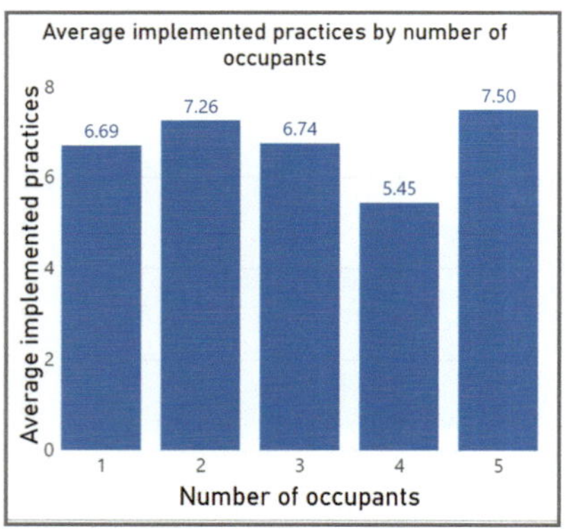

Figure 2.18 Influence of number of occupants.

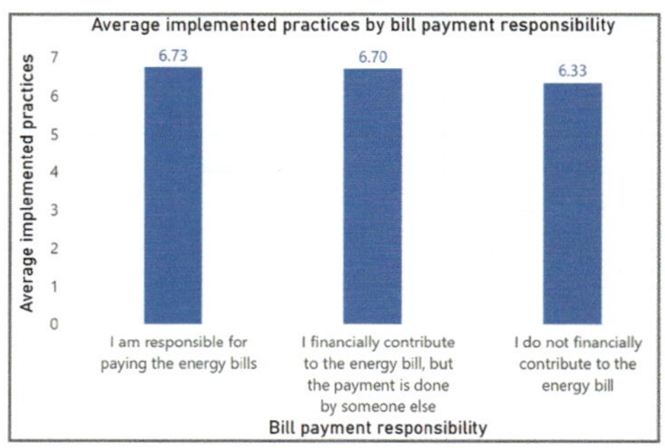

Figure 2.19 Influence of bill payment responsibility.

The number of people living in the household does not seem to significantly affect the adoption rate of energy-saving practices, as can be seen from Figure 2.18. The same applies to whether the respondent is responsible for paying the bill, not responsible but financially contributing, or not financially contributing at all (Figure 2.19).

The monthly average income of a household, on the other hand, shows a clear linear relationship with the energy-saving implementation. As can be seen from Figure 2.20, households in the lowest income group have by far the highest adoption rate. With growing income, the average adoption rate decreases. Significant correlation with adoption rates can also be observed

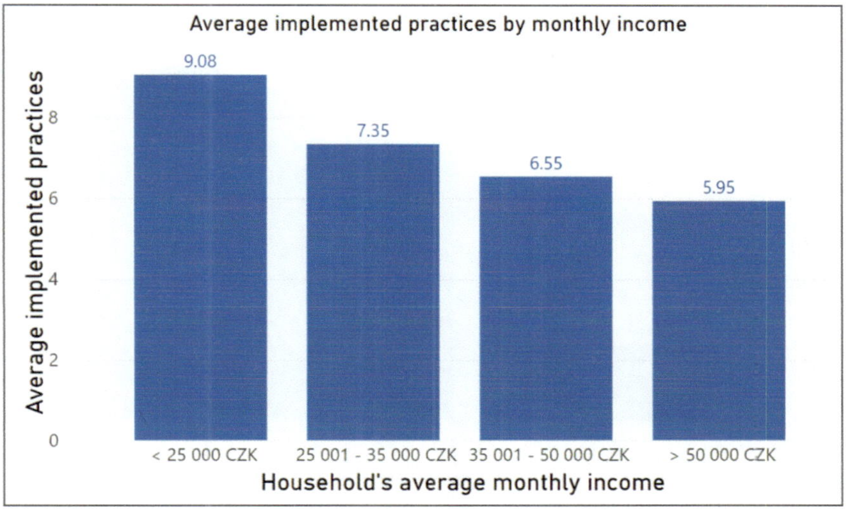

Figure 2.20 Influence of income.

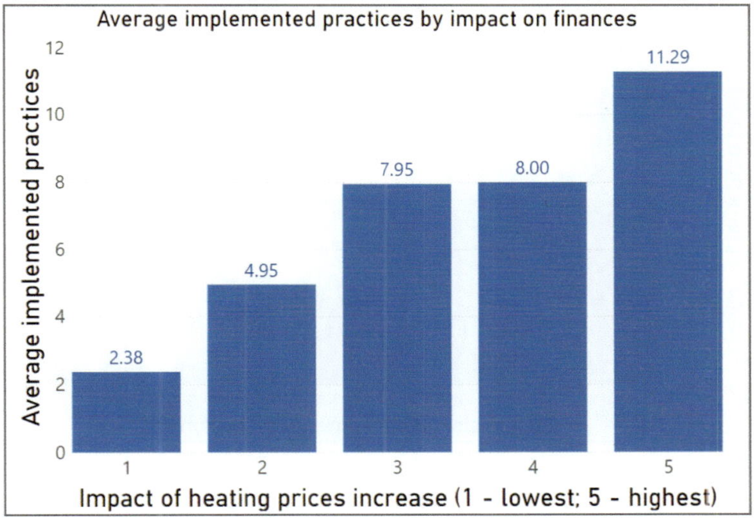

Figure 2.21 Influence of financial impact (HE).

when it comes to the impact of increased energy prices on household finances (Figures 2.21 and 2.22). The higher the impact of the energy prices, the more energy-saving practices respondents implemented.

In the case of the six statements, few correlations with the adoption rates can be observed. The first statement related to the interest in energy saving seems to confirm that people who agreed that their interest in energy-saving

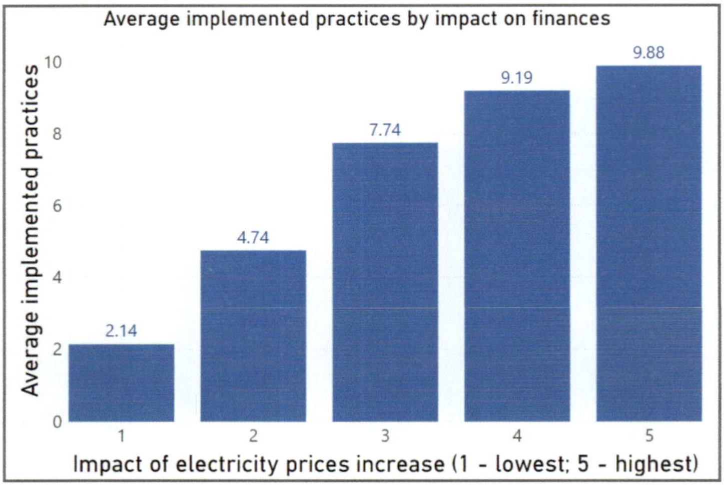

Figure 2.22 Influence of financial impact (EL).

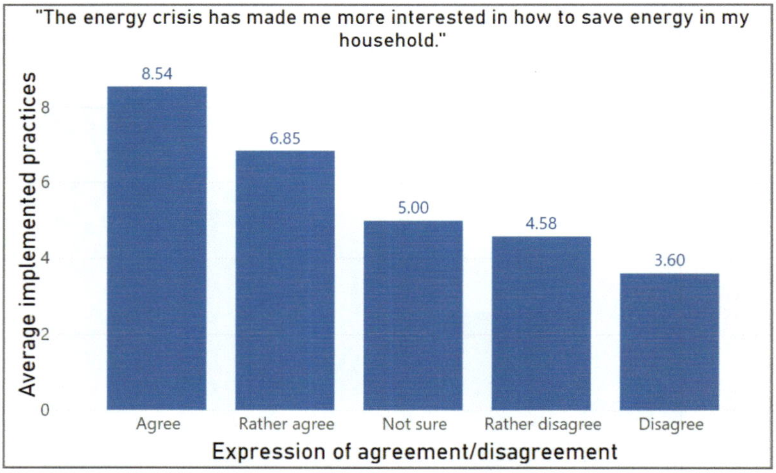

Figure 2.23 Influence of statement 1.

increased also adopted most practices since the crisis started. The less the respondents agreed with the statement, the fewer practices they adopted (Figure 2.23).

As for the second statement related to energy consumption awareness (Figure 2.24), the results indicate that respondents who became more aware of their energy consumption implemented more energy-saving practices than respondents who did not start tracking their energy consumption. However, respondents who were "not sure" whether their energy consumption awareness improved or not also reported a higher adoption rate.

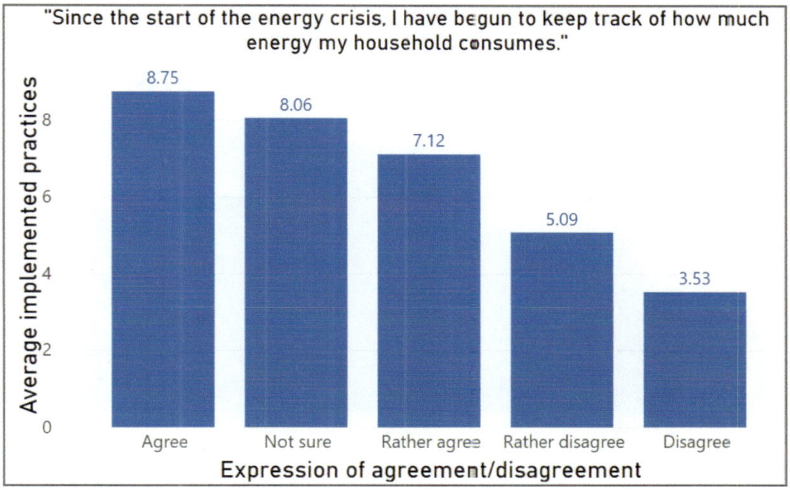

Figure 2.24 Influence of statement 2.

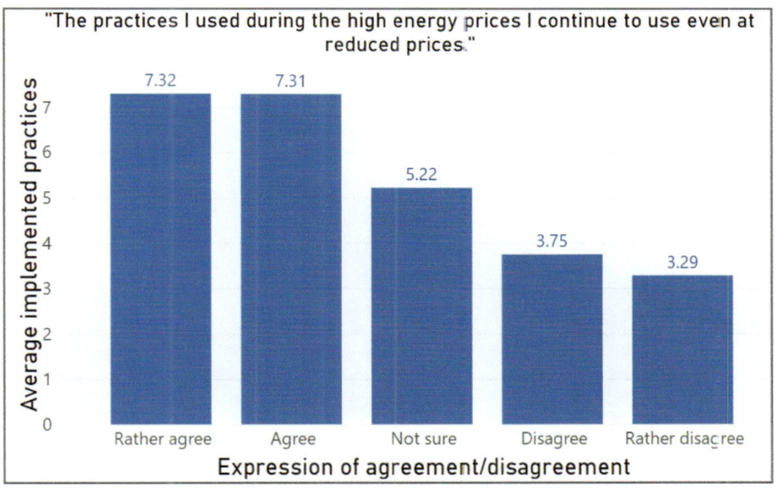

Figure 2.25 Influence of statement 3.

The third statement related to the will to continue with the energy-saving practices even after energy prices stabilize (Figure 2.25) shows that respondents who agreed or rather agreed with continuing have implemented more practices than those who disagreed or rather disagreed. However, the relationship is not linear. The fourth statement related to further potential for energy-saving (Figure 2.26) does not correlate with the adoption rate at all.

In the case of the fifth statement related to the role of financial motivation (Figure 2.27), it can be observed that respondents who agree, rather

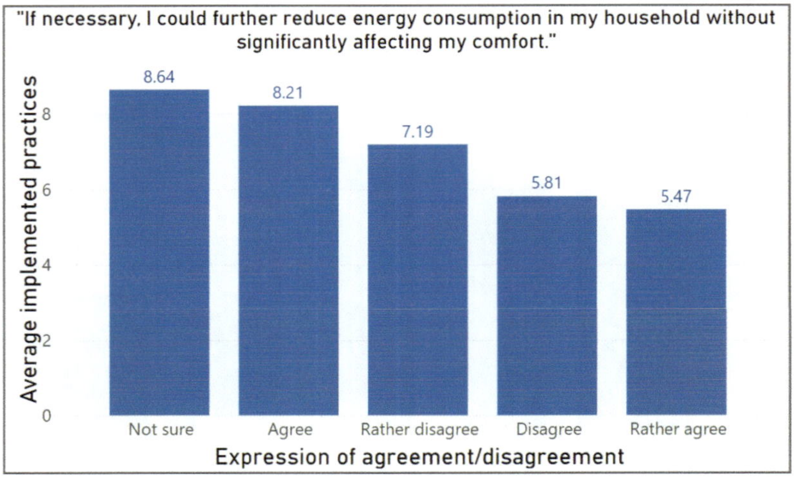

Figure 2.26 Influence of statement 4.

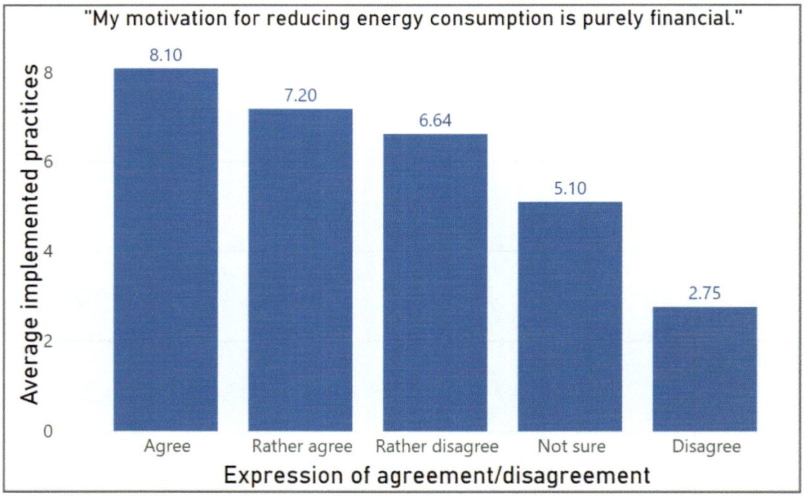

Figure 2.27 Influence of statement 5.

agree, or rather disagree have a relatively high adoption rate with not so significant differences, while people who disagree have a really low adoption rate. This might reflect that respondents who are motivated not only by financial motives (who disagree) have already been using many of the energy-saving practices before the prices increased, while the rest started only when they were financially affected by the crisis. Lastly, the sixth statement related to recognition of the positive impact of reducing household energy consumption on the environment (Figure 2.28) does not exhibit any correlation.

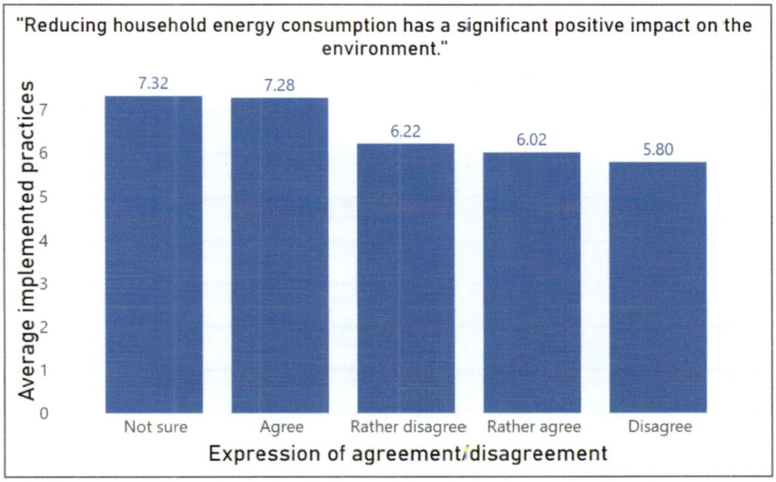

Figure 2.28 Influence of statement 6.

2.5 Discussion

If the results about the most commonly used practices are compared with the results from Kola-Kola-Bezka & Leki (2024), who investigated energy-saving behavior during the energy crisis in nearby Poland, certain similarities can be observed. For example, practices such as turning off the devices when not actively used, replacing lightbulbs with more efficient ones, or using laundry machines more efficiently rank in the top places. However, Kola-Bezka & Leki (2024) generally report much higher adoption rates, which might be caused by the fact that they focused on peripheral fossil-dependent regions that were particularly affected by the energy crisis. Moreover, the occurrence of heat insulation retrofitting was observed much more in their study. The whole comparison overview of the practices that were found in both research papers can be seen in Figure 2.29.

In comparison to findings from Liobikienė et al. (2023), who reported that only half of their respondents in Lithuania think that their individual energy consumption affects the overall consumption and subsequently the environment, result of this paper state that in Czech households 31% agrees and 38% rather agrees that their energy consumption ultimately has an impact on the environment. This reflects the positive trend in awareness of the general public of environmental considerations. An improvement is seen also in energy consumption awareness and energy-saving literacy, and even though Brounen et al. (2013) argue that energy literacy and awareness do not directly influence energy-saving, in the case of Czech households, there is a positive correlation.

Figure 2.29 Comparison of practices in the Czech Republic and Poland. (Source: Author; Bezka & Leki, 2024)

Even though there are a number of research papers stating that gender does not influence the energy saving behavior, results of this paper support the work of authors such as Räty & Carlsson-Kanyama (2010) or Belaïd & Joumni (2020), who argue that women are more likely to engage in energy saving. While some studies neglect the influence of age on energy-saving behavior (Hori et al., 2013), studies by Trotta (2018) and Kola-Bezka & Leki (2024) observed higher adoption rates among respondents of higher age. This paper observes mixed results, as there is no linear dependency, but the adoption rate is the highest in the case of respondents above 55 years old, followed by respondents in the 26–35 years age group. Education level is said to be an uninfluential parameter (Sardianou, 2007; Trotta, 2018), and in the results of this research, increasing education level does not correlate with a higher adoption rate; it is rather linked to specific education level groups. As for the average monthly income, results contradict Hori et al. (2013), who argue that income does not influence energy-saving and instead go in line with Kola-Bezka & Leki (2024), showing a higher occurrence of the energy-saving behavior change among lower income groups.

Recognition of the positive impact on the environment by saving energy was investigated in this paper as a potentially influential parameter to see whether it has the same effect as pro-environmental behavior (which has been proven to positively correlate with energy-saving). (Hori et al., 2013; Trotta, 2018) However, only awareness of the opportunity to make a positive impact does not seem to be enough for adoption of energy-saving behavior, and this variable did not turn out to be influential.

2.5.1 Delimitations

The influence of the field of employment on energy-saving behavior was eventually not assessed, as the question in the survey offered too many options, which resulted in some employment options only having a few respondents, so the sample would not be representative. In general, research could have included more respondents to provide a statistically more accurate overview.

When analysing the influence of the individual variables on energy-saving behavior, only one independent variable and one dependent variable were considered at the time. However, energy-saving behavior is possibly influenced by several variables at the same time, so multiple linear regression could be applied for data analysis to provide a more complex overview.

2.6 Conclusion

This study explored the impact of the 2022 energy crisis on the adoption of low-tech and no-tech energy-saving behaviors among Czech households. Through a detailed survey, it aimed to identify which specific practices were most commonly adopted and to understand the demographic and socio-economic factors influencing these behaviors.

The findings reveal that Czech households responded to the energy crisis by significantly adjusting their energy consumption habits. The most commonly adopted practices included lowering household temperatures during unoccupied periods, using appliances like washing machines and dishwashers more efficiently, and turning off electronics when not in use. These low-tech and no-tech solutions were straightforward, practical, and immediately implementable, demonstrating that households can effectively respond to energy crises through behavioral changes.

The analysis also highlighted several key factors influencing energy-saving behaviors. Age, sex, and income level were found to significantly affect the adoption rate of these practices. Older respondents and those with lower incomes were more likely to implement energy-saving measures. Women showed a higher tendency to engage in energy-saving behaviors compared to men. Additionally, households severely impacted by the increased energy prices were more proactive in adopting energy-saving measures, underscoring the role of financial motivation in behavior change.

The study's results also indicated an improvement in awareness and interest in energy-saving practices among Czech households. A significant number of respondents expressed a willingness to continue their energy-saving behaviors even after energy prices stabilize, suggesting a potential for long-term shifts in energy consumption habits.

However, despite the positive trends in energy-saving behavior, the study also identified areas for improvement. For example, while many households adopted basic energy-saving practices, fewer invested in more substantial measures like retrofitting or upgrading to more efficient appliances. This might indicate a need for policies that not only promote immediate energy-saving behaviors but also support households in making longer-term investments in energy efficiency.

In conclusion, this research contributes to the broader discourse on sustainable living practices and energy policy by providing evidence-based insights into household energy-saving behaviors during a crisis. Policymakers, energy providers, and community leaders can leverage these findings to design targeted interventions that support households in enhancing their energy efficiency, both in times of crisis and under normal conditions. Future research could expand on these findings by exploring the combined influence of multiple socio-demographic factors on energy-saving behaviors and by including a larger and more diverse sample of respondents to increase the generalizability of the results.

References

Alexander, S., & Yacoumis, P. (2018). Degrowth, energy descent, and 'low-tech'living: Potential pathways for increased resilience in times of crisis. *Journal of cleaner production*, *197*, 1840–1848.

Belaïd, F., & Joumni, H. (2020). Behavioural attitudes towards energy saving: Empirical evidence from France. *Energy Policy*, *140*, 111406.

Brounen, D., Kok, N., & Quigley, J. M. (2013). Energy literacy, awareness, and conservation behaviour of residential households. *Energy Economics*, *38*, 42–50.

Energetický regulační úřad. (2023a). *Domácnosti v roce 2022 rekordně šetřily elektrinou.* Retrieved May 13, 2024, from https://eru.gov.cz/domacnosti-v-roce-2022-rekordne-setrily-elektrinou

Energetický regulační úřad. (2023b). *Loňská spotřeba plynu byla nejnižší za posledních 8 let, meziročně se snížila o pětinu.* Retrieved May 13, 2024, from https://eru.gov.cz/lonska-spotreba-plynu-byla-nejnizsi-za-poslednich-8-let-mezirocne-se-snizila-o-petinu

Guan, Y., Yan, J., Shan, Y., Zhou, Y., Hang, Y., Li, R., ... & Hubacek, K. (2023). Burden of the global energy price crisis on households. *Nature Energy*, *8*(3), 304–316.

Hall, N., Romanach, L., Cook, S., & Meikle, S. (2013). Increasing energy-saving actions in low income households to achieve sustainability. *Sustainability*, *5*(11), 4561–4577.

Havlík, V., & Kluknavská, A. (2023). Our people first (again)! The impact of the Russia-Ukraine War on the populist Radical Right in the Czech Republic. *The impacts of the Russian invasion of Ukraine on right-wing populism in Europe*, 91.

Hoffmann, A., Nanaki, E., Enevoldsen, P., & Xydis, G. (2023). A behavioral change study in Denmark engaging car drivers in reducing fuel consumption: The key is in the message. International Journal of Sustainable Transportation, 17(2), 118–127.

Hori, S., Kondo, K., Nogata, D., & Ben, H. (2013). The determinants of household energy-saving behaviour: Survey and comparison in five major Asian cities. *Energy Policy*, *52*, 354–362.

Jing, X. (2023). The Russia-Ukraine War and Energy Security: Impact and Policies, From a European Perspective. *Highlights in Business, Economics and Management*, *3*, 215–222.

Kola-Bezka, M., & Leki, K. (2024). Household energy behaviour in the times of crisis: Lessons for policy initiatives from peripheral, fossil-dependent regions of the European Union. *Energy Policy*, *188*, 114113.

Koumoulides, D., Katsenios, N., Kasimatis, C. N., Xydis, G., & Efthimiadou, A. (2022). Socio-economic impact of the imposed lockdowns in food chains: a case study in Cyprus. Environments, 9(11), 137.

Liobikienė, G., Matiiuk, Y., & Krikštolaitis, R. (2023). The concern about main crises such as the Covid-19 pandemic, the war in Ukraine, and climate change's impact on energy-saving behaviour. *Energy Policy, 180*, 113678.

Martinsson, J., Lundqvist, L. J., & Sundström, A. (2011). Energy saving in Swedish households. The (relative) importance of environmental attitudes. *Energy policy, 39*(9), 5182–5191.

Ozili, P. K., & Ozen, E. (2023). Global energy crisis: impact on the global economy. *The Impact of Climate Change and Sustainability Standards on the Insurance Market*, 439–454.

Räty, R., & Carlsson-Kanyama, A. (2010). Energy consumption by gender in some European countries. *Energy policy, 38*(1), 646–649.

Sardianou, E. (2007). Estimating energy conservation patterns of Greek households. *Energy policy, 35*(7), 3778–3791.

Tin, T. T. H., & Xydis, G. (2025). The power of battery energy storage systems: unlocking home efficiency. International Journal of Emerging Electric Power Systems, (0).

Tóth, G., Jáger, V., Kovalszky, Z., Bóday, P., Dénes, Á., Kincses, Á., & Szép, T. (2023) Characteristics of Household Energy Consumption in the Shadow of the Russia-Ukraine War-a Case Study from Hungary. *Available at SSRN 4408442*.

Trotta, G. (2018). Factors affecting energy-saving behaviours and energy efficiency investments in British households. *Energy policy, 114*, 529–539.

Zhou, X. Y., Lu, G., Xu, Z., Yan, X., Khu, S. T., Yang, J., & Zhao, J. (2023). Influence of Russia-Ukraine war on the global energy and food security. *Resources, Conservation and Recycling, 188*, 106657.

Appendix 1 – Chapter 2

1. How old are you?
   (1) ❍ 18–25 years
   (2) ❍ 26–35 years
   (3) ❍ 36–45 years
   (4) ❍ 46–55 years
   (5) ❍ 56 years and over

2. What is your gender?
   (1) ❍ Male
   (2) ❍ Female
   (3) ❍ Other

3. What is your current employment status and in which sector do you work?
   (1) ❍ Student
   (2) ❍ Employed – Agriculture, Forestry, and Fishing (including all activities related to cultivation, extraction, and breeding)
   (3) ❍ Employed – Mining Industry and Energy (raw material extraction, energy production, and distribution)
   (4) ❍ Employed – Manufacturing Industry (all manufacturing from light to heavy industry)
   (5) ❍ Employed – Construction and Engineering (includes everything from construction to architectural design and urban planning)
   (6) ❍ Employed – Trade and Services (retail, wholesale, gastronomy, and accommodation)
   (9) ❍ Employed – Transport and Logistics (everything from operating transportation to managing infrastructure)
   (10) ❍ Employed – Information and Communication Technologies (ICT) (software development, data management, telecommunications)
   (11) ❍ Employed – Financial Services and Real Estate (banking, insurance, investments, property management)

(12) ❍ Employed – Professional, scientific, and technical services (legal services, accounting, consulting, research, and development)

(13) ❍ Employed – Public administration, defense, and social services (public administration, military services, social work, public health)

(7) ❍ Unemployed

(8) ❍ Retired

4. What is your highest level of education?

(1) ❍ Primary education

(2) ❍ Secondary education without graduation

(3) ❍ Secondary education with graduation

(4) ❍ Higher education (bachelor's degree)

(5) ❍ Higher education (master's degree)

(6) ❍ Postgraduate education

5. Select the type of your residence:

(1) ❍ Detached family house

(3) ❍ Semi-detached or terraced house

(2) ❍ Apartment in an apartment building

(4) ❍ Mobile home or caravan

(5) ❍ Other

6. Do you own or rent your residence?

(1) ❍ Owned

(2) ❍ Rented

7. How many people, including yourself, live in your household?

(1) ❍ 1 person

(2) ❍ 2 people

(3) ❍ 3 people

(4) ❍ 4 people

(5) ❍ 5 or more people

8. Do you contribute to or are responsible for paying energy bills in your household?
   (1) ❍ I am responsible for paying the bills
   (2) ❍ I contribute financially but someone else handles the payment
   (3) ❍ I do not contribute financially

9. What is the average monthly net income of your household?
   (1) ❍ 0–25,000 CZK
   (2) ❍ 25,001–35,000 CZK
   (3) ❍ 35,001–50,000 CZK
   (4) ❍ More than 50,000 CZK

10. What type of heating does your residence use?
    (1) ❍ Gas boiler
    (2) ❍ Electric boiler
    (4) ❍ Solid fuel boiler (coal, wood, etc.)
    (3) ❍ Heat pump
    (5) ❍ Electric heater
    (6) ❍ Other _____

11. What do you use for water heating in your residence?
    (1) ❍ Gas water heater
    (2) ❍ Electric water heater
    (4) ❍ Solid fuel water heater (coal, wood, etc.)
    (3) ❍ Heat pump
    (5) ❍ Solar panels
    (6) ❍ Other _____

12. To what extent has the increase in heating costs affected your household finances?
    (1) ❍ 1 (Not at all)
    (2) ❍ 2
    (3) ❍ 3
    (4) ❍ 4
    (5) ❍ 5 (Critically)

13. To what extent has the increase in electricity costs affected your household finances?
    (1) ❍ 1 (Not at all)
    (2) ❍ 2
    (3) ❍ 3
    (4) ❍ 4
    (5) ❍ 5 (Critically)

14. Which of the following household appliances do you have in your residence? (Check all that apply)
    (1) ❑ Refrigerator
    (2) ❑ Freezer
    (3) ❑ Dishwasher
    (4) ❑ Washing machine
    (5) ❑ Dryer
    (6) ❑ Television
    (7) ❑ Computer
    (8) ❑ Electric stove (classic, ceramic, induction)
    (9) ❑ Gas stove
    (10) ❑ Electric oven
    (11) ❑ Gas oven
    (12) ❑ Microwave oven
    (13) ❑ Air conditioning

15. Heating – Which of these measures did you use in your household to reduce energy consumption after the increase in energy prices? (Check all that apply)
    (1) ❑ Lower the temperature in rooms used without needing to wear warmer clothing
    (2) ❑ Lower the temperature in rooms where you need to wear warmer clothing
    (3) ❑ Not heating rarely used rooms
    (4) ❑ Lowering the temperature during the day when the residence is not used

(5) ❑ Lowering the temperature at night

(6) ❑ Additional insulation of the residence

(7) ❑ Replacing/sealing doors and windows to reduce heat loss

(8) ❑ Replacing the heating device with a new one to increase efficiency and reduce energy costs

(9) ❑ Other _____

16. Personal hygiene – Which of these measures did you use in your household to reduce energy consumption after the increase in energy prices? (Check all that apply)

(1) ❑ Bathing/showering less frequently

(2) ❑ Taking a shower instead of a bath

(3) ❑ Using cooler/cold water

(4) ❑ Shortening the duration of showers

(5) ❑ Installing aerators/low-flow showerheads

(6) ❑ Other _____

17. Dish washing and laundry – Which of these measures did you use in your household to reduce energy consumption after the increase in energy prices? (Check all that apply)

(1) ❑ Running the dishwasher less frequently/only running the dishwasher when more than 90% full

(2) ❑ Washing dishes manually instead of using a dishwasher

(3) ❑ Lowering the water temperature when washing dishes manually

(4) ❑ Running the washing machine less frequently/only running the washing machine when more than 90% full

(5) ❑ Lowering the water temperature when washing laundry/shortening the washing time

(6) ❑ Washing laundry manually instead of using an automatic washing machine

(7) ❑ Running the dryer less frequently/only running the dryer when more than 90% full

(8) ❑ Using a drying rack instead of an automatic dryer

(9) ❑ Replacing the washing machine/dishwasher/dryer with a new one to increase efficiency and reduce energy costs

(10) ❑ Other _____

18. Other household appliances – Which of these measures did you use in your household to reduce energy consumption after the increase in energy prices? (Check all that apply)

(1) ❑ Increasing the temperature in the refrigerator/freezer

(2) ❑ Turning off the television when no one is actively watching

(3) ❑ Turning off the computer when no one is actively using it

(4) ❑ Replacing installed bulbs with more energy-efficient alternatives

(5) ❑ Not using electric lighting during daylight hours

(6) ❑ Unplugging household appliances (microwave, television, etc.) when not in use

(7) ❑ Adjusting cooking habits to save energy (planning meal preparation and cooking in larger quantities)

(8) ❑ Other _____

For each of the following statements, indicate how much you agree:

19. "The energy crisis has made me more interested in how to save energy in my household."

(1) ❍ Disagree

(2) ❍ Rather disagree

(3) ❍ Not sure

(4) ❍ Rather agree

(5) ❍ Agree

20. "Since the start of the energy crisis, I have begun to keep track of how much energy my household consumes."

(1) ❍ Disagree

(2) ❍ Rather disagree

(3) ❍ Not sure

(4) ❍ Rather agree

(5) ❍ Agree

21. "The practices I used during high energy prices, I continue to use even when prices are lower."
    (1) ❍ Disagree
    (2) ❍ Rather disagree
    (3) ❍ Not sure
    (4) ❍ Rather agree
    (5) ❍ Agree

22. "If necessary, I could further reduce energy consumption in my household without significantly disrupting my comfort."
    (1) ❍ Disagree
    (2) ❍ Rather disagree
    (3) ❍ Not sure
    (4) ❍ Rather agree
    (5) ❍ Agree

23. "My motivation to reduce energy consumption is purely financial."
    (1) ❍ Disagree
    (2) ❍ Rather disagree
    (3) ❍ Not sure
    (4) ❍ Rather agree
    (5) ❍ Agree

24. "Reducing household energy consumption has a significant positive impact on the environment."
    (1) ❍ Disagree
    (2) ❍ Rather disagree
    (3) ❍ Not sure
    (4) ❍ Rather agree
    (5) ❍ Agree

3

Assessing the Grid Impact of Electric Vehicles in Denmark: Challenges, Solutions, and Policy Implications

Vivien Vanda Kuti[1], George Xydis[1,2]

[1]Department of Business Development and Technology, Aarhus University, Denmark
[2]Department of Mechanical Engineering, University of the Peloponnese, Greece

Abstract

This paper examines the effect of the growing electric vehicle (EV) adoption on Denmark's electricity grid with a focus on grid stability, renewable energy integration, and policy implications. The understanding of the interaction between EVs and the electricity grid helps to reach the goal that Denmark is aiming for, which is to be independent of fossil fuels by 2050. This research uses a mixed-methods approach, combining quantitative data analysis from reliable sources and qualitative insights from existing literature. Key findings include the significant growth in EV adoption supported by government incentives and policies. Furthermore, EVs offer opportunities for grid stability through vehicle-to-grid (V2G) technology, which can help mitigate the intermittency of renewable energy sources (RES). However, the increased electricity demand from EVs necessitates grid upgrades and the use of smart charging strategies. The paper also emphasizes the need for extending the policy and regulatory framework to facilitate EV integration, highlighting continuous improvements in the charging infrastructure. The chapter concludes with a discussion of challenges, solutions, and policy actions, including optimizing smart charging strategies and leveraging emerging technologies to enhance grid stability. The transition to electric vehicles presents a significant opportunity to enhance grid stability and integrate renewable energy sources

DOI: 10.1201/9788743808923-3

effectively in Denmark. This chapter provides insights and practical recommendations for policymakers, transmission operators, and stakeholders to support Denmark's ambitious energy transition goals.

3.1 Introduction

EVs are becoming more and more popular as technology and the infrastructure advance. EVs are considered the transportation of the future as the population is moving towards reducing the transportation-related oil and gas consumption and carbon emission to slow down global warming and increase energy independence. Furthermore, EVs can store electric energy and could possibly mitigate the intermittency of renewable energy sources such as wind and solar energy, using V2G technology.

To have EVs that are using fossil fuel-free energy from the grid, the countries need to adapt and utilise their renewable energy sources. This chaterwill discuss opportunities for renewable energy injection into the grid from EV battery storage, with Denmark serving as a great example. The Danish government has set the target that Denmark should become independent of fossil fuels by 2050, with high emphasis on the transportation sector. (*Promotion of Electric Vehicles EU Incentives & Measures Seen in A Danish Context*, n.d.)

Denmark's electricity grid will be the focus of this paper, given that Denmark has 80% of renewable energy integrated into its energy mix. (*Denmark – Countries & Regions – IEA*, n.d.) Denmark is also a leading developer of renewable energy technologies such as wind and solar parks, and power-to-x operations. Wind power accounts for more than half of Denmark's electricity generation (more approaching 75% lately) (Baral & Xydis, 2024). To manage the intermittency of wind power, Denmark currently uses solar energy and fossil resources to provide a stable electricity flow to its residents and industries. (*Denmark – Countries & Regions – IEA*, n.d.)

The number of EVs in Denmark has exponentially increased in recent years, and it is expected that more consumers will opt for EVs over traditional gasoline or diesel-powered cars. The Danish government is trying to urge car owners to buy or switch to EVs, to meet the European Union's (EU) 2030 climate goals for reducing CO_2 emissions. They are doing this by reducing taxes for electric car owners and reducing the number of gasoline and diesel cars sold in the country. Furthermore, Denmark has built an infrastructure that covers the needs for public charging and provides price benefits for charging stations and special parking places exclusively for EVs, and has further investments planned for the increasing demand in the upcoming years.

This chapter is going to focus on vehicles that are described as electric vehicles. The types of vehicles used for the research are the plug-in hybrid electric vehicles (PHEV), and battery electric vehicles (BEV). The other types of vehicles are excluded from the research, as they do not have the availability of electricity storage from the grid. In later sections, the PHEVs and BEVs are going to be referred to as EVs.

This study aims to assess the grid impact of EVs in Denmark, with a specific focus on understanding the challenges, solutions, and policy implications associated with EV integration into the electricity grid. By analysing EV penetration, charging behavior patterns, and grid interaction dynamics, this chapter seeks to provide insights into the broader implications of EV adoption for grid stability, reliability, and renewable energy integration. Additionally, the research aims to explore potential strategies and policy interventions to address grid integration challenges, optimize energy infrastructure, and support the transition to electric mobility in Denmark.

3.2 Literature Review

Electric vehicles (EVs) have emerged as a promising solution to address the challenges of transportation-related emissions and the need for energy independence. The literature review of this chapter consists of academic articles found in the topic of electric vehicles and their impact on grid stability.

3.2.1 Grid impact of EVs

Previous studies have investigated the grid impact of EVs, focusing on factors such as charging behavior, demand patterns, and grid stability considerations. The literature review research written by Richardson (2013) provides a very good overview of the literature on the topic and sheds light on the areas where future research is needed. His research highlights EV's potential to support grid stability, reduce peak demand, and impact infrastructure requirements through V2G technology and advanced control strategies. (Richardson, 2013) These findings emphasize the opportunities in EV integration to the grid to enhance stability and efficiency.

3.2.2 Integration of renewable energy sources

On top of the grid impact, EVs also have the potential to support the integration of renewable energy sources by serving as flexible grid assets and storage devices. Research by Pillai (2010) emphasises the role of V2G systems for achieving stable and efficient power systems with high renewable energy penetration. The article by Dallinger & Wietschel (2012) also concludes that plug-in electric vehicles have significant potential to balance intermittent renewable energy sources. (Dallinger & Wietschel, 2012; Pillai, 2010) This potential for balancing intermittent renewable energy highlights the need for supportive policy and regulations to facilitate EV integration.

3.2.3 Adoption factors and policy

Policy and regulatory frameworks play a crucial role in shaping the grid impact of EVs and helping their integration into the electricity grid. A study by Haustein et al. (2021) examined how people in Denmark and Sweden are accepting and using BEVs. The study concluded that in Denmark, there are more barriers that affect EV adoption. Another study by Noel et al. (2018) proves the same ideology. This study highlights that the EV penetration in Denmark is below the optimal level, and it is mainly due to the high cost difference between EVs and internal combustion engine vehicles. However, it is mentioned that as EV penetration grows, the policies will need to be improved to enhance electrification (Haustein et al., 2021; Noel et al., 2018). Addressing these policies and adoption factors are essential to maximize the benefits of EVs.

3.2.4 Gaps and future research

Despite the significant advancements in understanding the impact of EVs on grid stability and renewable energy integration, several gaps remain that need further investigation. Richardson, (2013) recommends further research to be done to explore the effectiveness of smart charging strategies, economic implications, and emerging technologies such as V2G integration. Haustein et al. (2021) suggest that Denmark needs to improve the infrastructure of the charging stations and emphasize marketing of electric cars to encourage adoption. The article by Noel et al. (2018) also advises that Denmark's policymakers will need to enhance the charging infrastructure and address the local grid impact as EV penetration will grow and reintroduce tax exemptions and benefits to accelerate EV integration. A study by Jayakrishnan R. Pillai (2010) suggests five potential areas for future research within EV market penetration, V2G technology, and grid impacts. (Haustein et al., 2021; Noel et al., 2018; Pillai, 2010; Richardson, 2013)

This research contributes to filling the identified gaps by focusing on the integration of EVs and V2G technologies to enhance grid stability and renewable energy use. By examining the economic impacts and infrastructure requirements, this study provides valuable insights for future policy implications. Additionally, understanding consumer behavior and adoption factors allows for the development of effective incentives and programs to promote EV use. Ultimately, this research aims to support the transition to a more sustainable and resilient energy system, leveraging the potential of EVs to balance intermittent renewable energy sources and improve grid efficiency.

3.3 Methodology

In this section, the methodologies used in the research paper are going to be described.

Research question:
How does the growing electric vehicle adoption affect the electricity grid stability and renewable energy integration in Denmark?

Objective:
The objective of this study is to assess the impact of electric vehicles (EVs) on grid stability in Denmark, with a focus on the challenges and solutions and the policy implications. A mixed-methods approach was adopted, combining quantitative data analysis and qualitative insights from research papers to provide a comprehensive understanding of the grid impact of EVs.

3.3.1 Data sources

The data for this study were sourced from various authoritative databases and reports. The data sources are considered secondary sources, as the data were gathered by external authorities and only used in the research for analysing and validating concepts.

- EV deployment: Data on the number of EV sales from January 2018 to April 2024 were obtained from Danmarks Statistik. Furthermore, data publications from the Danish Transmission System Operator (TSO), Energinet, and the Danish Ministry of Transport were used.
- Most sold EV models: The table of most sold EV types included in the research was taken from an online article. The data validity of this source is questionable; however, the article sourced the data table from Bilstatisktik.dk, which is considered a reliable source.

- Daily average electricity demand: The data for the daily average electricity demand was calculated from a dataset derived from the datasets of Energi Data Service. The calculation was based on the kWh hourly consumption from January 2024 to April 2024.
- Use of IT in the Danish population: The data was found on Danmarks Statistik's website. It is based on interviews that Danmarks Statistik has conducted in 2020.
- Policy: The information on current policies was taken from reliable authorities such as the International Energy Agency (IEA) and the Danish Energy Agency.

Furthermore, other data were found from published academic articles that are mentioned in the literature review and the analysis of this paper. To ensure a comprehensive understanding of the current state of knowledge, a systematic literature review was conducted. The primary databases used for the literature search included Google Scholar and ScienceDirect. Key search terms included "electric vehicles Denmark," "EV grid impact," "V2G technology," "renewable energy integration Denmark," and "smart charging."

3.3.2 Limitations

Even though the data was gathered carefully, some limitations need to be addressed.

For the analysis of EV deployment, multiple data sources and publications were reviewed, which showed similar outcomes but slightly different numbers. Therefore, the most reliable source was chosen to be included in the paper.

The data for most EV types sold could not be directly sourced from Bilstatisktik, as it is not publicly available.

The analysis used certain assumptions and simplifications in modelling the daily electricity consumption of EVs. These assumptions may not fully capture the complexities and variability inherent in real-world driving conditions, charging infrastructure availability, and user preferences. Furthermore, the calculation of all mentioned vehicle types would have been beneficial for the outcome of the calculation, but due to the limited character number, it was chosen to be simplified. The analysis did not account for certain external factors that could influence EV electricity consumption, such as changes in weather patterns, traffic congestion, road conditions, and driver behavior. These factors could have a significant impact on daily consumption but were not explicitly addressed in the research.

The research focused on a specific time period for data collection and analysis, which may limit the generalizability of the findings over longer time spans or across different seasons. Variations in electricity consumption patterns throughout the year were not fully explored.

The findings of the study are specific to the context of Denmark and may not be directly applicable to other regions with different electricity grid structures, energy policies, and EV adoption rates. Cross-country comparisons were beyond the scope of this research.

3.4 Analysis and Findings

This section of the research is going to focus on analysing the EV deployments and trends in Denmark, assessing the grid impacts of EVs and the potential use of EVs in renewable energy integration.

3.4.1 EV deployment in Denmark

Denmark has seen an increase in EV adoption by the public in recent years. The growing adoption is supported by government policies, an increase in the public's environmental awareness, and advancements in EV technology.

Figure 3.1 shows the sum of EV sales in Denmark from January 2018 to April 2024. The graph shows an overall trend of increase. By April 2024, the growth in BEV sales reached 23,4872 registered vehicles, compared to January 2018, where only 8781. The sales of PHEVs have also seen a rise through the years, though they are much smaller than BEVs. According to the publication of Energinet, the sales of EVs are going to rise to 1.5 million by 2035 in Denmark (Schmidt et al., n.d.).

The growth of EV sales in Denmark necessitates the corresponding expansion of EV charging infrastructure. The number of charging stations increased by 8100 in 2023, totalling 17,400 charging stations across the country, with extra attention to fast and ultra-fast chargers. Though the country's fast chargers still only make up 13% of the total, ultra-fast chargers take up 9%. The public charging network is built and operated by the charging operators, the most dominant being Spirii, Clever, and EON in 2023 (Ministry of Transport, 2024).

The Danish government has plans to further support the transition to EVs, including the further decrease of new diesel and petrol car sales by 2030, and enhance low-emission zones (Danish Ministry of Climate, 2019). Furthermore, the government has set the goal to reach 760,000 EVs by 2030, which seems viable if the growth continues at the current rate.

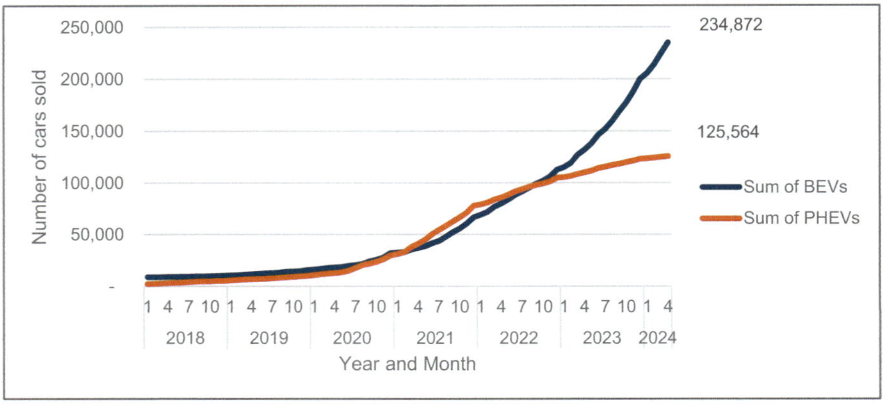

Figure 3.1 Number of EVs sold based on historical data, (Bestand Af Motorkøretøjer Efter Område, Køretøjstype, Brugerforhold Og Drivmiddel – Statistikbanken – Data Og Tal, n.d.) graph is own production.

3.4.2 Grid interaction and stability

EV's interaction with the grid presents both opportunities and challenges for grid stability. As the number of EVs in Denmark is expected to grow, assessing their impact on the grid is essential for determining the electricity supply.

The charging habits of EV users largely affect the work of the Danish Transmission System Operator (TSO), Energinet's daily operation. The work of the TSO is to balance the demand and supply of energy in the country and provide a stable flow of energy to the end-users and prevent blackouts (Schmidt et al., n.d.).

EV users can select to charge their cars at public charging points or have chargers installed at their home. While the public charging demand is much harder to estimate, for EV charging at home, the EV owners can choose their own charging plan from their electricity provider. A simple charge plan is a system where the vehicles start charging as soon as they are connected to the charger. The delayed charge plan offsets the charging of the EV for a certain time, for example, to charge during the night, when the electricity is at the lowest cost. The most efficient way of charging is smart charging, which uses some kind of IoT[1] device to detect the most beneficial time of charging, which could be based on the low cost of energy, low demand, or excess energy produced (Richardson, 2013). In Denmark, currently to charge

[1] Internet of Things

Table 3.2 Average consumption of the top 5 EV types

No.	Company	Model	Market share (%)	Average consumption (kWh/100 km)	Source of information
1	Tesla	Model Y	13.20%	18	(*Køretøjets Strømforbrug | Tesla Support Danmark*, n.d.)
2	Tesla	Model 3	13.04%	16	(*Køretøjets Strømforbrug | Tesla Support Danmark*, n.d.)
3	Volkswagen	ID.4	5.14%	20	(*Volkswagen ID.4 Pro Performance (2021–2023) Price and Specifications – EV Database*, n.d.)
4	Skoda	Enyaq iV	4.95%	19	(*Škoda Enyaq iV – Consumption WLTP | EVspecs*, n.d.)
5	Volkswagen	ID.3	3.67%	17	(*Volkswagen ID.3 Pro (2021–2022) Price and Specifications – EV Database*, n.d.)

the EV at home, the owner needs to purchase or rent a charging box from an electricity provider. Some providers, such as Spirii, Andrel Energi, Circle K, and NRGi provide the opportunity to download their app, which monitors the user's electricity usage and gives recommendations on when to charge the car based on the electricity price.

The load of charging also depends on the type of EV and its battery capacity. In Denmark, the most sold EVs include Tesla, Volkswagen, and Skoda (Appendix 2).

To be able to relate the charging of the type of cars to the electric grid demand, it is necessary to calculate the consumption of the EV types and calculate an average consumption per day for all cars sold based on the market share. In Table 3.2, the average consumption can be seen for each car type. Only the top five car types are going to be calculated to simplify the tables.

Next, the number of EVs for each model is going to be calculated based on the number of EVs sold in 2024, which is 360,436 and their market share percentage. Then, it is assumed that each car drives 50 km per day on average, and the total consumption is calculated based on that. The numbers are calculated with the following formula:

Table 3.3 Number of EV models sold and their consumption.

No.	Company	Model	Market share	Average consumption (kWh/100 km)	Estimated cars	Energy consumption (kWh)
1	Tesla	Model Y	13.20%	18	47,570	428,130
2	Tesla	Model 3	13.04%	16	46,987	375,896
3	Volkswagen	ID.4	5.14%	20	18,526	185,260
4	Skoda	Enyaq iV	4.95%	19	17,828	169,386
5	Volkswagen	ID.3	3.67%	17	13,221	112,379

$$\text{Daily energy consumption (kWh)} = (\text{Average consumption (kWh/100 km)} \times 50\text{ km})/100 \times \text{Number of cars}$$

The following table includes the calculated number of cars based on the market share and includes the estimated energy consumption per day.

Just the top five types of EVs consume 1,271,051 kWh of energy together per day. Compared to the average daily total electricity consumption in Denmark, it is taking up close to 1% of the total (*Energi Data Service | Datasets | Consumption Coverage, Hourly National Statistics*, n.d.). Though this number might seem small, it should be considered that the calculation only regards 40% of the EVs, and the distance driven is only an assumed number. A lot of factors affect the daily consumption of each EV owner, such as driving distance, driving style, vehicle efficiency, charging habits, battery size, charging efficiency, etc. Therefore, this number is only an indication, and the total consumption is expected to be higher.

Research from Statistics Denmark shows that in Denmark, 36% of the citizens use some type of IoT home appliance in 2020, but only 10% of them use devices to measure their electricity, water, and heat consumption. (*NYT: Danske Hjem Bliver Mere Intelligente – Danmarks Statistik*, n.d.) This can lead to the assumption that only a small percentage of the citizens use the available smart charging devices for their EVs. As the number of EVs is expected to grow, the unmanaged charging can have a destabilizing effect on the grid, causing voltage fluctuations and power losses (Energinet, 2022).

3.4.3 Renewable energy integration

The integration of renewable energy sources (RES) into the electricity grid is a crucial part of reaching the energy goals in Denmark. The EVs can have a significant impact on helping the transition to increased renewable energy reliance in the country.

EVs, because of the battery, can be considered as mobile energy storage. EVs through the charging process can store energy and could potentially be used to store excess energy produced during peak production periods. This is most useful for storing renewable energy, as the renewable energy sources are highly intermittent. The smart charging technologies could essentially help the owners determine the optimal charging time to match peak production. (Richardson, 2013)

EVs could also be used to provide electricity back to the grid using V2G technology. This could potentially stabilize the electrical grid and optimize RES, as EVs could provide the energy to the grid in times of low production. In Denmark, where the adoption of EVs is increasing, the potential for V2G technology to support renewable energy integration is substantial. In variable renewable energy-dominated regions, such as Denmark, V2G is one of the most cost-effective flexibility options for optimizing energy system operations.(Nagel et al., 2024)

The large-scale adoption of V2G technology could significantly decrease price differences between power regions and reduce price variability, leading to a flatter price duration curve. This would result in more stable electricity prices and enhance the economic efficiency of the energy system. Nagel et al. (2024) found that V2G can supply up to 11% of system load during peak electricity price periods, and as much as 55% in specific regions like DK2 in Denmark. This contribution helps in mitigating the intermittency of renewable energy sources.

While large-scale V2G adoption may lead to increased charging prices for end-users due to the reduction of very low-price periods, it also results in reduced peak prices, which counteract the increased charging costs. The overall effect is an improvement in system efficiency and a reduction in system emissions.

3.4.4 Discussion

The analysis of EV deployment trends in Denmark reveals a significant increase in EV adoption in recent years, driven by supportive government policies, rising environmental awareness, and technological advancements. However, this rapid growth presents both challenges and opportunities for various stakeholders involved in the transition to EVs.

3.4.5 Challenges and solutions

The increasing number of EVs presents challenges for grid stability and reliability, especially during peak charging times. This can lead to voltage fluctuations and potential power outages. The potential solution would be to encourage the use of smart charging and demand response programs, which could help manage the load peak. Encouraging off-peak charging and using time-of-use pricing could distribute the demand more evenly throughout the day.

The current infrastructure will need further development to keep up with the growing number of EVs. There will be a need for more charging stations, particularly fast and ultra-fast chargers. Denmark has extended plans to install further public charging stations and has funds to install chargers in housing associations as well (Ministry of Transport, 2024).

The initial cost of EVs and charging infrastructure can be excessive for some consumers. Additionally, there may be resistance to adopting new technologies and changing charging habits. Financial incentives such as subsidies, tax refunds, and low-interest loans could lower the economic barriers. Public awareness campaigns and educational programs can help increase acceptance and adoption of EVs and their technologies (Haustein et al., 2021).

3.4.6 Policy implications

To address these challenges and facilitate the integration of EVs into Denmark's energy system, several policy measures should be considered.

Denmark has current policies that incentivise EV adoption. Denmark has introduced reduced registration taxes for EVs, tax reimbursement for electric company cars, and reduced prices for electricity when charging an EV. Additionally, the Danish government is planning to phase out the sale of new diesel and petrol cars by 2030 and to create low-emission zones in urban areas to encourage the use of cleaner vehicles. (*Ny Aftale: Billigere Elfirmabiler Og Skattefri Opladning På Arbejdspladsen Skatteministeriet*, 2023)

While Denmark has made significant steps in promoting EV adoption and infrastructure development, there are further policy implications to consider for ensuring the continued growth and integration of EVs into the energy system.

The government should provide targeted aid and funding to support the development of both public and private charging infrastructure. This includes increasing the number of fast and ultra-fast chargers, particularly in areas with high demand. Collaborations between the government and private sector entities could accelerate the development of the charging infrastructure.

Promoting the adoption of smart charging systems and V2G technology is essential. Financial incentives for the installation of smart chargers and V2G-compatible vehicles can encourage owners to adopt these technologies. Further policies should support the integration of EV charging infrastructure with renewable energy sources to solve the intermittency issue with V2G. Offering renewable energy certificates for EV owners who charge their vehicles using renewable energy sources could further promote the use of green energy and support Denmark's renewable energy goals.

Public awareness and education campaigns are critical for increasing acceptance and adoption of EVs, smart charging, and V2G technology. Furthermore, encouraging employers to provide workplace charging facilities and offering tax incentives for businesses that invest in EV charging infrastructure can enhance the convenience of owning an EV.

Finally, establishing systems for ongoing data collection and analysis will help monitor the impact of EVs on the grid and the effectiveness of policies and support. This data could suggest future policy adjustments and ensure that strategies remain relevant and effective. Regularly engaging with stakeholders, including owners and potential buyers, industry representatives, and energy providers, will ensure that policies are aligned with the needs and capabilities of all parties involved. This collaborative approach could help innovation and address challenges.

3.5 Conclusion

The integration of electric vehicles into Denmark's electricity grid presents both significant challenges and opportunities. By leveraging smart charging and V2G technologies, enhancing infrastructure, and implementing supportive policies, Denmark can optimize the benefits of EVs to enhance grid stability and renewable energy integration. Addressing economic barriers through financial support and expanding public awareness can further accelerate the transition to EVs. Continuous monitoring and stakeholder engagement will be essential in developing strategies and ensuring a resilient and sustainable energy system.

The insights gained from this study contribute to a deeper understanding of the grid impact of EVs and provide valuable recommendations for policymakers, utilities, and stakeholders. As Denmark continues its journey towards a sustainable energy future, the successful integration of EVs will play a crucial role in achieving its climate and energy goals, ultimately contributing to a more flexible and sustainable energy system.

References

Baral, S., & Xydis, G. (2024). Unleashing the economic potential of wind power for ancillary services. International Journal of Emerging Electric Power Systems.

Dallinger, D., & Wietschel, M. (2012). Grid integration of intermittent renewable energy sources using price-responsive plug-in electric vehicles. *Renewable and Sustainable Energy Reviews*, *16*(5), 3370–3382. https://doi.org/10.1016/j.rser.2012.02.019

Danish Ministry of Climate, E. and U. (2019). *Denmark's Integrated National Energy and Climate Plan under the Regulation of the European Parliament and of the Council on the Governance of the Energy Union and Climate Action.*

Denmark - Countries & Regions - IEA. (n.d.). Retrieved May 9, 2024, from https://www.iea.org/countries/denmark

Energi Data Service | Datasets | Consumption Coverage, Hourly National Statistics. (n.d.). Retrieved May 26, 2024, from https://www.energidataservice.dk/tso-electricity/ConsumptionCoverageNationalDecl

Energinet. (2022). *Vehicle grid integration research rethinking energy international alignment.*

Haustein, S., Jensen, A. F., & Cherchi, E. (2021). Battery electric vehicle adoption in Denmark and Sweden: Recent changes, related factors and policy implications. *Energy Policy*, *149*, 112096. https://doi.org/10.1016/J.ENPOL.2020.112096

Køretøjets strømforbrug | Tesla Support Danmark. (n.d.). Retrieved May 25, 2024, from https://www.tesla.com/da_dk/support/power-consumption

Ministry of Transport. (2024). *The number of fast and quick charging stations is multiplied | Ministry of Transport.* https://via.ritzau.dk/pressemeddelelse/13775696/antallet-af-hurtig-og-lynladestandere-mangedobles?publisherId=5278762&lang=da

Nagel, N. O., Jåstad, E. O., & Martinsen, T. (2024). The grid benefits of vehicle-to-grid in Norway and Denmark: An analysis of home- and public parking potentials. *Energy*, *293*, 130729. https://doi.org/10.1016/j.energy.2024.130729

Noel, L., Zarazua de Rubens, G., & Sovacool, B. K. (2018). Optimizing innovation, carbon and health in transport: Assessing socially optimal electric mobility and vehicle-to-grid pathways in Denmark. *Energy*, *153*, 628–637. https://doi.org/10.1016/J.ENERGY.2018.04.076

Ny aftale: Billigere elfirmabiler og skattefri opladning på arbejdspladsen Skatteministeriet. (2023). https://skm.dk/aktuelt/

presse-nyheder/pressemeddelelser/ny-aftale-billigere-elfirmabiler-og-skattefri-opladning-paa-arbejdspladsen

NYT: Danske hjem bliver mere intelligente - Danmarks Statistik. (n.d.). Retrieved May 26, 2024, from https://www.dst.dk/da/Statistik/nyheder-analyser-publ/nyt/NytHtml?cid=41576

Olsen Lasse. (2024). *Tal om elbiler – statistik og salgstal i Danmark (april 2024)*. https://mobilsiden.dk/nyheder/opkoblede-biler/tal-om-elbiler-statistik-og-salgstal-i-danmark/

Pillai, J. R. (2010). *Electric Vehicle Based Battery Storages for Large Scale Wind Power Integration in Denmark*.

Promotion of electric vehicles EU Incentives & Measures Seen in A Danish Context. (n.d.). Retrieved May 9, 2024, from www.eaea.dk

Richardson, D. B. (2013). Electric vehicles and the electric grid: A review of modeling approaches, Impacts, and renewable energy integration. *Renewable and Sustainable Energy Reviews*, *19*, 247–254. https://doi.org/10.1016/j.rser.2012.11.042

Schmidt, L., Graduate, A., Datahub, E., & Andersen, M. H. (n.d.). *Driving Towards Grid Balance An Architecture Clarification White Paper Energinet | Tonne Kjaersvej 65, 7000 Fredericia*.

Škoda Enyaq iV - Consumption WLTP | EVspecs. (n.d.). Retrieved May 25, 2024, from https://www.evspecs.org/technical-data/skoda/enyaq-iv/consumption

Vehicle to Grid (V2G) Technology - IEEE Innovation at Work. (n.d.). Retrieved May 28, 2024, from https://innovationatwork.ieee.org/vehicle-to-grid-v2g-technology/

Volkswagen ID.3 Pro (2021-2022) price and specifications - EV Database. (n.d.). Retrieved May 25, 2024, from https://ev-database.org/car/1531/Volkswagen-ID3-Pro

Volkswagen ID.4 Pro Performance (2021-2023) price and specifications - EV Database. (n.d.). Retrieved May 25, 2024, from https://ev-database.org/car/1314/Volkswagen-ID4-Pro-Performance

Appendix 2 – Chapter 3

Table A2.1 The most purchased EVs in Denmark (Olsen Lasse, 2024).

No.	Company	Model	Market share
1	Tesla	Model Y	13.20%
2	Tesla	Model 3	13.04%
3	Volkswagen	ID.4	5.14%
4	Skoda	Enyaq iV	4.95%
5	Volkswagen	ID.3	3.67%
6	Audi	Q4 e-tron	3.48%
7	Hyundai	Kona Electric	2.77%
8	Polestar	2	2.33%
9	Volvo	EX40	2.31%
10	Ford	Mustang Mach-E	2.25%
11	Tesla	Model S	2.20%
12	Renault	Zoe	2.18%
13	Kia	Niro EV	2.10%
14	Mercedes-Benz	EQA	1.91%
15	BMW	iX3	1.82%
16	Volkswagen	e-Golf	1.80%
17	Volkswagen	ID.5	1.71%
18	Cupra	Children	1.53%
19	Audi	e-tron	1.48%
20	Hyundai	Ioniq 5	1.46%
21	Nissan	Leaf	1.38%
22	Volkswagen	up!	1.35%
23	Renault	Megane E-Tech	1.31%
24	Kia	EV6	1.31%
25	BMW	i4	1.25%
26	BMW	i3	1.23%
27	Mercedes-Benz	EQB	1.10%
28	Mini	Cooper	0.98%
29	Fiat	500e	0.94%
30	Peugeot	e-208	0.94%
31	Others		16.85%
Total			**100%**

4

Public EV Charging: Comparing Spot and Fixed Electricity Pricing

Prabakar Paramasivan and George Xydis

Department of Business Development and Technology, Aarhus University, Denmark

Abstract

This research paper conducts a comparative analysis of spot and fixed electricity pricing models for electric vehicle (EV) chargers. Using data from EV charging providers, the study compares both pricing models in the context of EV charging. The findings shed light on challenges and opportunities from the transition of fixed to spot pricing in promoting EV adoption, optimizing charging infrastructure utilization, and supporting grid stability. Additionally, the paper explores how spot pricing can incentivize demand flexibility among EV users, potentially reducing peak-load stress on power systems while balancing cost-effectiveness for consumers. The insights derived from this analysis aim to inform policymakers, utility providers, and charging operators in designing pricing strategies that align with sustainable energy goals.

4.1 Introduction

Energy demand and supply are fundamental components of the modern energy market. There is a complicated relationship between demand and supply in the energy market. Nowadays, energy market is in a constant quest for grid stability, efficiency, and sustainability (Ucal & Xydis, 2020). Electric vehicles (EVs) are the future of sustainable transportation. As governments, universities, industries, and consumers increasingly prioritize environmental concerns, EVs have emerged as an important way to reduce carbon emissions and reduce reliance on non-renewable fossil fuels. A reliable public

DOI: 10.1201/9788743808923-4

EV charging infrastructure is required for a transition to EVs. This paper will focus on the context of fixed vs spot pricing for public charging of electric vehicles in Denmark.

4.2 Background Information

In this section, more detailed information about the context of the research work will be presented. In recent years, EV sales have increased in Denmark. EV charging infrastructure plays a vital role in the adoption of EVs. In 2023, 36% of the new cars sold in Denmark were Electric (*Fdm*, 2024), and 20% of used cars available for sale are electric (*Elbilviden.Dk*, 2024). The increase in EV sales is propelled by ambitious sustainability goals, conscious buying, and government initiatives like reduced tax for electric vehicles. The development of public EV charging infrastructure plays a crucial role in owning an EV car. Both normal charging (usually charging speed 22 kW per hour) and fast charging (charging speed greater than 100 kW per hour) are essential components of this infrastructure. This research will focus specifically on normal charging due to its importance for EV owners, particularly those residing in urban areas. Home charging is an option for EV users who have access to their own charger. This research will not focus on this because home charging is usually based on spot pricing.

In Denmark, public EV charging is offered by companies such as Clever, Eon, Spirii, etc. These companies have chargers located across Denmark. Users can charge their EV by having an account with these companies and initiate charging either through the mobile app of the respective companies or a charging brick issued by these companies. After using the charger, the user will be billed based on the energy consumed. This is how a normal charging process works. These companies usually charge a fixed price for electricity measured in kW charged by EVs. Figure 4.1 shows the pricing of the Clever and Eon in Denmark (*Clever.Dk*, n.d.).

Public EV charging companies have priced it around 3.5 DKK for 1 kW of electricity (*Eon*, n.d.). This research will use this price for comparison against spot electricity prices. Companies like Clever and Eon's purpose is to provide EV charging as a service to its users. They invest in EV charging infrastructure and try to be profitable in providing this service (Xydis & Nanaki, 2015). They are not involved in electricity production and distribution. This elaborate background information's purpose is to provide an extensive overview of the project's focus area within the context of public EV charging in Denmark.

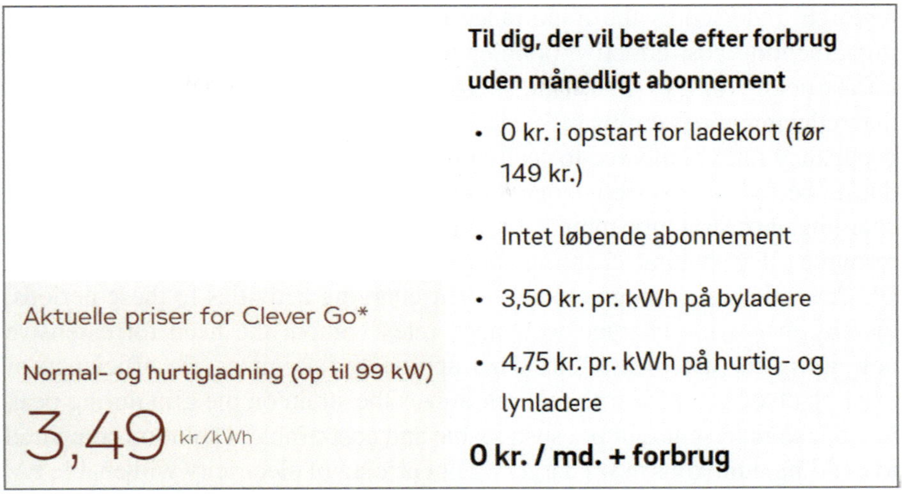

Figure 4.1 Public EV Charging pricing (Clever, Eon)

4.3 Literature Review

A major barrier hindering the growth of the EV market is the insufficient public charging infrastructure. Research in this area has shown that the adoption of EVs provides significant benefits in reducing carbon emissions. The total number of public charging stations and charging units in the world had reached 190 thousand and 1.26 million, respectively, by the end of 2015. One of the key factors affecting the development of charging infrastructures is the profitability of public charging infrastructures. Profitability depends on factors with direct impacts like charging demand, charging price, unit subsidy for construction, unit subsidy for operation, number of charging units, unit construction cost of charging units, ground rent, maintenance cost, electricity price, and location (Zhang et al., 2018). Among these factors, the pricing of electricity for the end user plays a significant role, as the end user is charged for the entire EV charging experience based on electricity consumption.

A study examines consumption rivalry and demand estimation in public charging infrastructure for EVs using novel real-time inventory data from a residential area of Amsterdam (Soetevent, 2024). Although consumption rivalry is largely a local issue confined to peak hours, the results indicate that expanding capacity solely to address this may not be cost-effective. Instead, policies focused on optimizing existing capacity – such as demand management strategies and incentives for off-peak charging – are recommended. This research underscores the importance of accurately estimating demand and implementing cost-effective policies to optimize the utilization of charging infrastructure for EVs (Pan et al., 2020). This paper (Soetevent, 2024) suggests that reducing prices during off-peak hours is likely to stimulate higher charger occupancy rates compared to scaling the capacity of chargers. Lower pricing during periods of reduced demand incentivizes EV owners to utilize existing charging infrastructure more effectively, optimizing the usage of available resources. By making charging more cost-effective during off-peak times, EV drivers are encouraged to shift their charging activities to these periods, thereby increasing charger occupancy rates without the need for extensive new infrastructure investment. This approach maximizes the efficiency of existing charging infrastructure, minimizes the strain on the grid during peak hours, and represents a more sustainable and cost-efficient solution compared to building numerous new chargers. Spot pricing of electricity will enable EV owners to look out for off-peak times to save EV charging cost.

Vehicle-to-Grid (V2G) technology enables EVs to interact with the electricity grid by allowing them to both charge from and discharge electricity back into the grid (Tin & Xydis, 2025). Essentially, it turns EV batteries into flexible energy storage units that can supply power to the grid during times of high demand or when renewable energy sources are not available (Womble & Xydis, 2021). V2G systems utilize bidirectional charging infrastructure to facilitate this two-way flow of electricity between EVs and the grid. EV users can potentially earn revenue by selling surplus energy stored in their vehicle batteries back to the grid during peak demand periods, while also helping to stabilize the grid. However, challenges such as infrastructure limitations, battery degradation, and low consumer awareness hinder V2G adoption (Kester, 2018).

4.4 Problem Statement and Research Questions

In Denmark, the transition to electric vehicles is happening at a fast rate, driven by environmental concerns and technological advancements. However, one important challenge lies in pricing electricity used in public EV charging. Electricity for domestic purposes is usually priced at the spot price. But the electricity used for EVs in public chargers is based on a fixed energy price. The debate between fixed and spot pricing models for EV charging has implications for consumer behavior, market dynamics, profitability of EV charging companies, and the overall sustainability of the transportation sector. Thus, there is a valid rationale to evaluate and compare the effectiveness of fixed and spot electricity pricing models in facilitating the widespread adoption of EVs.

4.4.1 Research questions

- Why do public EV charging companies prefer fixed electricity pricing over spot electricity pricing?
- What are the challenges and opportunities associated with transitioning from fixed to spot pricing for public EV charging?

4.4.2 Delimitations

This study, conducted using data from Denmark, aims to compare the spot and fixed electricity pricing models for EVs. The rationale for this comparison stems from the fact that the monthly electricity consumption of a typical two-person household in Denmark ranges from 100–150 kWh (*Enerdata*, n.d.). Electric consumers in Denmark have the option to choose between spot and fixed pricing based on the contract they establish with their energy provider. Considering that the average size of an EV battery is 60–85 kWh, charging an EV twice can equate to the energy consumption of a household over one month. Thus, extending pricing flexibility to EV users is deemed crucial in this research. Certain assumptions in the spot pricing model are made using past spot prices. Also, this research does not focus on home charging, where the EV user owns an EV charger and has the possibility of establishing a spot electricity contract with the electricity provider. Not every EV user can have access to their own EV charger, and the dependence on public EV charging is an important aspect of EV ownership. V2G in the context of electricity pricing is discussed; however, the important aspect of V2G adoption is battery degradation. Battery degradation of EVs by V2G is not the focus of the research.

4.5 Methodology Overview

This chapter outlines the systematic approach used in the study, serving as a foundation for addressing the research questions and problem statement. It details the methodology employed, including the rationale behind key decisions in data selection and analysis.

The methodology is divided into two key components:

- Data assessment (Bryman, 2014) – covering data selection, collection, and analysis.
- Trustworthiness assessment (Shenton, 2004) – evaluating reliability through qualitative and quantitative data validation.

4.6 Data Assessment

This section describes the criteria for data selection and its relevance to the study. The chosen data aligns with the research objectives, primarily focusing on qualitative sources due to their suitability for the analysis. Additionally, secondary data from credible institutions (e.g., Energinet, EV charging providers) was incorporated. While these sources were not originally intended to address the research questions directly, they provided valuable contextual insights. The following sections further explain the selection of both qualitative and quantitative data.

4.6.1 Qualitative data analysis

The qualitative research approach was selected for its ability to deliver in-depth insights into:

- Business models of public EV charging operators
- Variations in public EV charging infrastructure

4.6.2 Quantitative data supplement

To enhance the reliability of our findings, quantitative data was incorporated, focusing on:

- Comparative pricing structures of Danish EV charging providers
- Historical energy price data from authoritative sources (Energinet, DSI)

4.6.3 Systematic data collection methodology

A structured data gathering process was implemented, utilizing author's tacit knowledge of EV charging companies in Denmark. Thorough desktop research of EV charging companies and their methods, a detailed comparison of fixed and spot electricity prices was done throughout the project.

4.6.4 Secondary data

This study incorporates secondary data – information not originally generated for this specific research but repurposed for analytical use. Examples include publicly available historical energy price records for Denmark, as well as peer-reviewed academic literature. To ensure rigorous analysis, the selected data was evaluated through relevant theoretical frameworks, particularly focusing on:

- Public EV charging infrastructure dynamics
- Electric vehicle user behavior patterns

4.6.5 Trustworthiness assessment

The reliability of both data sources and analytical outcomes is critically examined in this section. By evaluating methodological choices and analytical tools, this assessment verifies result accuracy. A scientifically grounded approach was prioritized throughout the research to uphold data integrity, with separate evaluations for qualitative and quantitative components.

4.6.6 Qualitative data evaluation

Credibility was ensured through a comprehensive review of all applied theories and methodologies. Data was acquired via systematic desktop research, with particular attention to:

- Transferability (Shenton, 2004):
  The degree to which findings can be generalized to comparable contexts (e.g., applicability to other EU nations' public EV charging systems may require supplementary investigation).

- Dependability:
  The consistency of results if replicated under identical methodological conditions. Given the use of documented corporate data from actual EV charging providers, findings demonstrate high replicability.

- Confirmability:
  Measures taken to minimize researcher bias in desktop research-derived data. The predominance of externally verified qualitative sources ensures objective outcomes.

4.6.7 Quantitative data evaluation

This analysis focuses on result consistency and measurement precision:

- Reliability:
  The likelihood of obtaining identical results upon repeating the data collection process. All quantitative metrics were sourced from authenticated public repositories.

- Validity (Shenton, 2004):
  The accuracy of measured concepts. While developed for EV infrastructure analysis, this methodological approach shows potential applicability to alternative renewable fuel systems (e.g., hydrogen charging networks).

4.7 Analysis

In this section, the focus will be on comparing fixed and spot electricity pricing and then applying the context of EV charging companies. From this comparison, the research question "*Why do public EV charging companies prefer fixed electricity pricing over spot electricity pricing?*" will be answered.

First, the basic structure of the electricity market will be explained. The structure of an electricity market typically involves four main physical layers - generation, transmission, distribution, and load.

Generation refers to the production of electricity from various energy sources. Transmission involves the transfer of electricity generated to local distribution networks. The distribution layer involves the delivery of electricity from the transmission system to end-users, such as homes, businesses, and industries. Load represents the demand for electricity from the end-users mentioned. In Denmark, the transmission system operator is Energinet (*Eneginet*, n.d.). Energy generation is done by various Energy project developers. The distribution layer is provided by the energy companies. Now, both spot and fixed electricity pricing will be explained.

4.7.1 Spot electricity pricing

Spot or variable electricity pricing refers to electricity rates that fluctuate based on current market conditions, such as supply and demand dynamics, source of electricity, weather conditions, grid stability, and in rare cases, external forces like war, political situation, etc. Usually, it is the day-ahead market that determines the spot price. The day-ahead electricity market facilitates next-day electricity transactions through a centralized auction system, where participants submit bids and offers based on expected supply and demand. Transactions are matched at a clearing price, and financial settlements are made accordingly. This market enhances grid reliability and market liquidity while providing transparency and efficiency in electricity procurement (*Nordpool*, n.d.).

4.7.2 Fixed electricity pricing

Fixed electricity pricing refers to a consistent rate for electricity consumption. This fixed rate remains constant regardless of fluctuations in the electricity

Table 4.1 Fixed vs spot assumptions.

In DKK	Fixed	Spot 1	Spot 2	Spot 3
Electricity	3.5	2.6	.6	4.2
+CAPEX + OPEX + profit (0.5 DKK)	3.5	3.1	1.1	4.7
+CAPEX + OPEX + profit (1 DKK)	3.5	3.6	1.6	5.2

prices. Usually, there is an agreement to pay for a predetermined rate for the electricity they consume during the agreed pricing period. Fixed electricity pricing offers stability and predictability, allowing customers to budget effectively and avoid surprises on their pricing. These public EV charging companies enter long-term contracts with electricity generators to secure a stable supply of electricity at already committed prices.

According to (*Statistics Denmark*, n.d.), the average price of electricity in the second half of 2023 was 2.6 DKK, and two other pricing scenarios are used from (*Eneginet*, n.d.), with electricity prices of 0.8 DKK and 4.2 DKK.

Since the CAPEX, OPEX, and profit margin of the EV charging infrastructure companies are not known and out of the scope of this research, two assumptions of 0.5 and 1 DKK are made. The purpose of this table is to show various scenarios of pricing based on the assumptions. The green price indicates the EV charging price is less than the fixed price of 3.5 DKK, and the red price indicates the EV charging price is higher than the fixed price. Public EV charging companies prefer charging their users a fixed price for the following reasons.

Recover their CAPEX and OPEX on EV infrastructure – public EV charging companies have invested huge amounts of money to establish and run EV charging infrastructure, and by having a fixed price, it will be easier for them to predict and offset their CAPEX & OPEX.

Profitability – by having a fixed price, they can set a high profit margin after deducting all their expenses. Then these companies add their profit margin to establish their price for the consumers.

Simplicity from customer perspective – fixed pricing offers EV owners predictability and simplicity in managing their charging costs. With a fixed price per kWh, customers know exactly how much they will pay each time they charge their vehicle, regardless of fluctuations in electricity prices or charging demand.

Service provider perspective – public EV charging provides a service to EV users, the freedom to charge at their own price. Meanwhile, providing household electricity can be classified as providing basic necessities like water and heat.

4.7.3 Discussion vs conclusion

In this section, the focus will be on the second research question, *"What are the challenges and opportunities associated with transitioning from fixed to spot pricing for public EV charging?"* and provide a conclusion to the research. Few charging stations with dynamic pricing are also present (*Better Energy*, 2023). This charging station was introduced by Better Energy, primarily a solar energy generation company. However, these charging stations are very less in number and operate as a standalone case. The purpose of highlighting this is to provide evidence that public EV charging is possible with a spot pricing model.

4.7.4 Opportunities for EV charging companies in a transition to spot pricing

Changing pricing from fixed to spot pricing for EV charging companies can serve as a catalyst for innovation, leading to a much more profitable margin for the EV charging companies. Currently, there is an option to purchase only renewable energy for households. The energy distribution companies offset the non-renewable energy used by buying green energy certificates. Similarly, the same can also be introduced for EV charging.

Charging companies can look for collaboration with EV manufacturers and governments to share the costs of establishing charging infrastructure, easing their capital expenditure burden. In return, EV manufacturers benefit from increased EV sales due to improved charging accessibility. Government can benefit from faster transition to sustainable mobility (*Cbc Canada*, n.d.). An example of this can be the case of Norway, where its government policies helped to scale the charging infrastructure. In Norway, free public EV charging is widespread, supported by government incentives and initiatives. This policy encourages EV adoption, reduces emissions, and promotes sustainable transportation.

Offering spot energy prices can contribute to improving grid stability as users tend to charge their EVs when prices are low, which in turn corresponds to lower grid load. This behavior helps to balance electricity demand and supply more effectively, reducing peak demand. By incentivizing consumers to shift their energy consumption to off-peak hours through spot pricing. This promotes more efficient utilization of grid resources and infrastructure. This not only enhances grid stability but also supports the consumption of renewable energy sources, which might be wasted due to low demand.

V2G technology, despite its innovative potential, is complicated to use with a fixed electricity pricing model. Spot pricing model allows EV owners to adjust charging and discharging based on real-time market prices, optimizing energy usage and potentially generating revenue. This benefits grid stability and encourages renewable energy integration. With spot pricing, EV users have flexibility in returning the electricity from their EV to the grid when the price is higher. As already stated, the capacity of an EV battery can store large amounts of electricity for regular home usage. By utilizing this technology, EV charging companies can also profit from offering this service to their customers. By providing end users with dynamic pricing options during periods of fluctuating electricity prices, EV charging companies can achieve profitability without taking substantial risks.

4.7.5 Challenges for EV charging companies in the transition to spot pricing

Companies often resist changing established business models. These EV charging companies will try to continue to maintain the status quo. Also, with the current infrastructure, it is difficult to pass on the current spot price to the customer. It is difficult to keep the EV charger displaying the spot price.

Transitioning to a new pricing model in the EV charging industry can be challenging for companies. Adapting to alternative pricing structures may disrupt established revenue streams and require adjustments to the operating model. The pricing will become complicated for the end user and making it difficult to budget EV charging.

EV users may face challenges in optimizing their charging schedules to take advantage of lower spot prices, as they must constantly monitor market conditions and adjust their charging behavior accordingly. As explained in Table 4.1, the price difference between spot prices and fixed prices may not be that huge. The price differential between spot and fixed pricing models may not always be substantial enough to justify significant changes in behavior for EV users. From the assumption scenario made, the potential cost savings from shifting charging behavior to exploit lower spot prices may not outweigh the inconvenience and effort required to constantly monitor market conditions and adjust charging schedules.

In the ongoing debate between fixed and spot electricity pricing for public EV charging, existing companies often favor maintaining the status quo, as it aligns with their established business models and reduces potential risk to their revenue streams. Change can introduce uncertainty and require significant adjustments to operational strategies, which may not be favored by incumbents who have invested in existing infrastructure and market positions.

However, it is important to recognize that the EV charging industry is continually evolving, driven by technological advancements, policy changes, and shifting consumer preferences. While fixed pricing offers stability and predictability, there may be opportunities for companies to explore alternative revenue streams that do not solely rely on traditional pricing models.

By embracing innovation and adopting customer-centric approaches, companies can not only adapt to changing market dynamics but also create value for end-users and contribute to the transition towards a more sustainable mobility. In this context, the debate between fixed and spot pricing should not be viewed as a binary choice, but rather as part of a broader discussion on how to optimize the EV charging market for the benefit of everyone in the EV charging value chain.

References

Better Energy. (2023). https://www.betterenergy.com/news/better-energy-s-ev-charging-park-aims-to-encourage-flexible-energy-consumption-through-affordable-and-transparent- pricing/

Bryman. (2014). *Business Research Methods*.

Cbc canada. (n.d.). https://www.cbc.ca/news/climate/canada-norway-evs-1.7092003

Clever.dk. (n.d.). https://clever.dk/

Elbilviden.dk. (2024). https://elbilviden.dk/

Eneginet. (n.d.). Energinet.dk

Enerdata. (n.d.). https://www.enerdata.net/estore/energy-market/denmark/

Eon. (n.d.). https://www.eon.dk/privat.html

Fdm. (2024). https://fdm.dk/

Kester, J. (2018). Promoting Vehicle to Grid (V2G) in the Nordic region_ Expert advice on policy mechanisms for accelerated diffusion. *Energy Policy*.

Nordpool. (n.d.). https://www.nordpoolgroup.com/en/the-power-market/Day-ahead-market/ Pan, L., Yao, E., Yang, Y., C Zhang, R. (2020). A location model for electric vehicle (EV) public

charging stations based on drivers' existing activities. *Sustainable Cities and Society*, *5S*, 102192. https://doi.org/10.1016/j.scs.2020.102192

Shenton. (2004). *Strategies for Ensuring Trustworthiness in* ualitative *Research Projects*.

Soetevent, A. R. (2024). I'd like to move it! The effect of consumption rivalry on demand estimation: Evidence from the EV public charging market. *Journal of Environmental*

Economics and Management, *124*, 102929. https://doi.org/10.1016/j.jeem.2024.102929

Statistics Denmark. (n.d.). https://www.dst.dk/en/Statistik/emner/miljoe-og-energi/energiforbrug-og-energipriser/energipriser

Ucal, M., & Xydis, G. (2020). Multidirectional relationship between energy resources, climate changes and sustainable development: Technoeconomic analysis. Sustainable Cities and Society, 60, 102210.

Womble, M. S., & Xydis, G. (2021). Energy Storage: A Practical Solution to Increase Wind Energy Integration in the US Electricity Sector. In Renewable Energy for Mitigating Climate Change (pp. 41-54). CRC Press.

Xydis, G., & Nanaki, E. (2015). Wind energy based electric vehicle charging stations sitting. A GIS/wind resource assessment approach. Challenges, 6(2), 258-270.

Zhang, Q., Li, H., Zhu, L., Campana, P. E., Lu, H., Wallin, F., C Sun, Q. (2018). Factors influencing the economics of public charging infrastructures for EV – A review. *Renewable and*

Sustainable Energy Reviews, *S4*, 500–509. https://doi.org/10.1016/j.rser.2018.06.022

5

Energy Communities in Poland: Enablers of the Country's Energy Transition or Niche Concepts?

Aleksandra Krolikowska[1] and Milosz Panasik[1]

[1]Department of Business Development and Technology, Aarhus University, Denmark

Abstract

As energy transition is an ongoing topic across many countries all around the world, looking into its enablers is a highly relevant topic. Energy communities can be defined as one of the options for that, where they can be described as citizen-driven enablers of renewable energy actions. As technological advancement is highly country-specific, so is the popularity of such communities. The below research aims at defining what the situation of individuals' participation within the Polish renewable energy sector looks like currently and what the actual long-term potential of Energy Communities is there. The research is based on comparison with Germany, as it is both a direct neighbor of Poland and a well-experienced leader within the energy cooperatives sector for many years now. Even though the countries cannot be compared one-to-one on all levels, the primary and secondary information included in the paper highlights gaps and best practices that should be incorporated in Polish energy communities for them to be successful.

5.1 Introduction

One of the key priorities for Europe in the XXI century is the battle against climate change. The main action topic included in the international agenda, which aims to make a difference from a global perspective, is energy transition towards less- and zero-emitting sources, as well as putting an end to the

DOI: 10.1201/9788743808923-5

linear and fossil fuel-driven economy. On the opposite side of the centralized, fossil-fuel, and big-scale energy infrastructure, a local, community-driven, and renewable innovation has been emerging as an interesting alternative.

Based on the adjusted European goals for 2050, which have been enclosed within the European Commission's Fit-for-55 package, Renewable Energy Directive 2018/2001/EU (RED II) raises the expected share of renewable energy to reach 32% by 2030 (European Commission, 2018). In order to push the member states towards that ambitious target, a set of mandatory requirements has been defined, encouraging them to revise or develop national plans relating to renewable energy communities and self-consumers. There, these general directives leverage the citizen-driven energy transition as a consistent part of the decarbonization process (European Commission, 2018).

The concept of energy communities has been known and practised for many years now in most European countries. What is more, literature agrees on the number of technological as well as social benefits relating to the framework. Yet, the split in the acceptance and popularity differs drastically from country to country, and the full-scale adaptation of the concept has not yet been visible in some nations, such as Polish society. Even though Poland does not differ significantly geographically from a country in which such communities are highly successful, namely its western neighbor Germany, the number of registered and operating communities is close to zero.

The research aims at identifying the current status of the adaptation of energy communities in the Polish market as well as defining the potential future development of the sector. One of the main research questions tries to assess whether such a concept can be a feasible enabler of the country's energy transition toward renewables, or is it barely a niche and temporary concept.

The identified literature concerning energy communities (EC) in Poland is not very extensive, but the available material is highly differentiated and discusses various topics around economic and technological feasibility, social community perception towards Renewable Energy Sources (RES), as well as political alienation of the government across the European arena.

5.1.1 Benefits

There is a common understanding that decentralized, citizen-driven energy transition may generate differentiated benefits to the regions as well as the national energy system. Some, but not limited to, include long term positive influence on grid balancing capabilities (Wawrzyniak et al., 2021), the local labour market benefits (Blazejowska, 2018; Furmankiewicz et al, 2021), increased regional innovation (Cichowska et al., 2021), environmental protection (Blazejowska, 2018), greater energy security and independence (Blazejowska, 2018; Wawrzyniak et al., 2021, Capellan-Perez, et al., 2020, Cichowska et al., 2021; Eiser et al., 2010), and developed citizen capital (Furmankiewicz et al., 2021).

The topic discussed naturally correlated with an environmentally centred socio-political discussion that is becoming more and more present within the academic literature in the country. There is a common understanding that the flexibility and social acceptance gained by the concept can become one of the key positive drivers of renewable energy (RE) transition in the country (Bomberg and McEwen, 2012; Wawrzyniak et al., 2021; Skoczkowski et al., 2016).

5.1.2 Barriers

Other than the benefits, literature strongly focuses on the barriers as well as the main enablers of the concept's implementation and the general growth of the RES industry. The most frequently discussed topic is the national governmental influence on policymaking and law, as well as its limited political will. Juszczyk et al. (2022) define political influence as a key element of the RES implementation, especially in a country like Poland, where the coal lobby still represents a strong political power. More holistically, Bomberg and McEwen (2012) point out a need for structural and symbolic resources in the community in order to create engagement and acceptance. Favorable political structures and their support of society are considered necessary in the transition period, where just the narrative of communication may have a significant impact on the process (Furmankiewicz et al., 2021). The focus on communication has been brought up as well in the context of education and awareness sharing, which should focus on the publicly available knowledge about the full potential for the local labour market development as well as the economic benefits available for the stakeholders (Skoczkowski et al., 2016; Juszczyk et al., 2022; Hielscher et al., 2022).

5.1.3 Drivers

No matter the influential position of the central government, the first movers in the collective energy industry have been kept on a rather local level or as external organizations. International support and political pressure have been identified by Furmankiewicz et al. (2021) and Lupi et al. (2021) as some of the initial drivers of communication. Hielscher et al. (2022) define the year 2017 as the beginning of a more active and aggressive narrative of social groups against passive government operations. According to the authors, Poland joined the international narrative towards increased RES development efforts by e.g., active participation in movements such as Earth Strike and Extinction Rebellion. However, based on Capellan-Perez et al. (2020), Polish initiatives are more driven by businesses than citizen movements and individuals. The main priorities of investment are energy savings and economic gains, rather than RES and social collaboration.

The strategic connection between individuals, enterprises, as well as central government has been recognized by some researchers in local administration. The regional government takes an active part in the development of RES investments, which have a major impact on the final national energy mix. Nevertheless, authors see that there is a limited number of studies and literature on the function of the members in regard to the RES (Drozdz et al., 2021). What displays a rather difficult relationship between the local organs are examples brought up by Hilscher et al. (2022), indicating that the Prime Minister's cabinet is more likely to support increased power over the administration, discredit bottom-up initiatives, as well as disregard local problems in the perspective of national opportunities.

5.1.4 Technologies

Decentralized and grid-connected RES infrastructure has been proven to be a feasible investment in rural communities in Poland (Krawczak, 2019). The majority of cases studied took solar PVs as the primary technology for tests. Nevertheless, the primary RE share is covered by onshore wind, which became the leading and rapidly growing technology back in 2010 and practically stagnated between 2016 and 2019 (Drozdz et al., 2021). Jasinski et al. (2021a, 2021b) analyse the effectiveness of energy cooperatives and conclude that wind energy can be twice as effective as solar. However, there is no single solution for community energy technology as local demand and supply characteristics differ every time. What is more, investing in RES technology brings rural areas to lower energy losses as well as decreases dependencies on the energy purchased from the grid.

5.1.5 Community

According to Furmankiewicz et al. (2021), due to the socialist system between 1945 and 1989, community renewable energy does not have a strong historical tradition in Poland, which makes it more difficult and less natural to develop such structures nowadays. Lupi et al. (2021) recognise the collective action initiatives as a relatively new social innovation concept in the country when compared with the rest of the Western European region. Authors do not find community cooperation social structures existing in the country, which makes such functions being carried out by other institutions and individuals.

Interestingly, Juszczyk et al. (2022) recognize that Poland and Nordic countries, with Finland as an example in the researched case, represent similar problems, constraints, and drivers in regard to the RES implementation. Even though the sale of problems and cultural contexts may be different, there is a strong connection with regard to the struggle within the energy transition process. One of the mentioned problems, also in other works such as at Skoczkowski et al. (2016), was a narrative of "Not in my backyard," which is characterized by a negative approach towards visible infrastructural investments in proximity to the individual. Maruszewska (2021) indicates that technological developments in the local community have a negative impact due to strong attachment to the traditionally owned land in rural communities and traditional landscapes, which has personal importance to the Polish rural nation. This feeling of land attachment has been valued over the connection to the local community as well as neighborly human relations. Additionally, the author claims that public opinion from the local communities' perspective has a constraining effect on decision making, as the residents would not decide to invest in a technological project that could be noticeable by others, since it could relate to unwanted judgment and becoming spotlighted.

Lastly, a positive perspective on the development of future RES is identified by Cichowska et al. (2021), as in the paper authors analyse youth perception of renewables. The students in the study agreed that such sources must become the key element of the future energy mix. The group that has been identified as more knowledgeable and willing to engage is the rural community. Nevertheless, no matter the increased social engagement, Capellan-Perez et al. (2020) claim that post-Soviet countries are still an under-represented region in literature in regard to the Community Renewable Energy topic.

5.2 Methodology

This section of the report has been covered by a review of the research design, methods, data collection, analysis, as well as limitations and delimitations.

5.2.1 Research design

In order to understand the topic of energy communities in general, but specifically in Poland as the chosen target country, it can be stated that the vast majority of information available can be categorized as qualitative. That includes official reports and presentations prepared by businesses, which include text and descriptions rather than numbers and figures, audio, and video recordings from various webinars, as well as documents and interviews conducted by the researchers.

One of the designs available when performing a qualitative type of research is grounded theory, which, in essence, comprises defining a theory based on participants' views. As there are multiple forms of qualitative data incorporated in the research, there are also multiple levels of data collection involved in the process, which align with the just-mentioned theory.

What is more, following the research design defined above, inductive reasoning was applied, as the primary goal of the research paper was to define whether the concept of energy communities in Poland is a niche idea or if there is actual potential for it to develop over time. Such an approach arises from observation as well as data and information collection on the topic, through their analysis and patterns' identification, and ends up with a determined theory.

5.2.2 Research method

From the various qualitative methods available for application, the research focused on interviews, image analysis, as well as document data. While interviews are considered primary data sources, as the information is obtained by the researcher directly from the source, the latter two fall under the secondary data category. Image analysis will be performed in the form of looking at countries' maps, both in terms of the current layout of ECs and also with regard to solar and wind potential across the lands. Document data will be incorporated in the form of reports and literature available on the topics of energy transition and energy communities.

5.2.3 Data collection

Subsequent to the research method is the data collection technique. As previously defined, the researchers have decided to conduct an interview with the initial purpose of obtaining first-hand information from an EC representative, viz business development partner from Eisall Energy. Considering the wide extent of the subject matter and type of chosen research topic, the questions asked during the interview were designed to remain open-ended to enable the collection of subjective and detailed perspectives of the participant on the discussed topic.

Aside from the interview, internet-based research was performed to acquire detailed and up-to-date information on ECs in general, their development in Poland, and potential perspectives for the future.

5.2.4 Data analysis

Considering that the below research incorporates a qualitative design, data analysis was performed along with other elements within the research. Data and information were gathered simultaneously. There was no pre-specified data that was supposed to be collected prior to data analysis and report structuring. On the contrary, some parts of the information gathered during the initial phase of research created an outline for the final report. Nevertheless, some additional findings from the interview conducted were added and analysed, while continuing with other parts of the research, which reshaped the report's structure along the way. What is more, the accuracy of the obtained information was verified during the entire project lifespan, enabling supplementary results to be included.

In terms of research's validity and reliability, those two aspects can be defined on opposite levels. Due to a rather high level of information cross-checking as well as looking into details and facts highly relevant to the defined topic and scope, the first factor, viz, validity, is set to be on a fairly high level. Reliability, on the other hand, as a direct result of the highly volatile market of RE and ECs, as well as the unpredictable nature of the current political situation in Poland in that regard, can be assessed to be on a rather low level.

5.2.5 Limitations and delimitations

Limitations: There were a few external factors outside of the researchers' scope that limited the research below in the following manner.

Firstly, even though there are currently 11 ECs established in Poland, only two of them have made it through all of the steps in the process and have become officially approved by the Polish National Agricultural Support Centre (KOWR). The nine remaining ECs are considered as of a significantly smaller scale and thus do not own an official website or contact information that could be used for research purposes. Therefore, the research was limited to the two ECs that made it through the entire process, as the researchers could not obtain additional information from the ones that were still underway. What is more, even though the two existing ECs had accessible contact information, the availability of interviewees was limited, which affected the number of details acquired during the interview.

Secondly, apart from data collection limitations, the researchers faced time limitations, as the project timeline was pre-defined by the university's curriculum and semester plan.

Delimitations: When it comes to delimitations, namely boundaries of the research study set by the researchers, there are three to mention.

As the topic of the paper suggests, the research's scope was initially limited to Poland as the target country. Nevertheless, to create a brief benchmark, Germany was included as the neighboring country with similar natural conditions necessary for the development of RE sources.

Closely related to that, the second aspect concerned focusing solely on energy communities in Poland and information available in that manner, rather than going into details on other available community-based concepts, such as energy clusters or prosumers. Therefore, the research below will not include a thorough analysis of other collective energy-related initiatives.

Lastly, due to the high domination of photovoltaics in the Polish RE market, followed by wind in significantly lower quantities, the research limited its scope to these two types of renewables.

5.3 Results

5.3.1 Natural conditions

Germany was chosen as the comparison country within this research as it is a direct neighbor of Poland, placed at nearly the same height and latitude. That means that this country has potentially similar natural conditions necessary for the development of the RE sector, such as wind and solar power. What needs to be emphasized here is to acknowledge that for the development of wind and solar farms within a country, many more conditions need to be fulfilled than simply having the resources available. These include both technical aspects related to the actual design of the products, but also country-specific economic, political, and environmental situations. In this part of the report, the focus is on comparing the potential of regions when looking solely at given wind speed and solar radiation across the two regions in Europe based on the visuals and information coming from global wind and solar atlases.

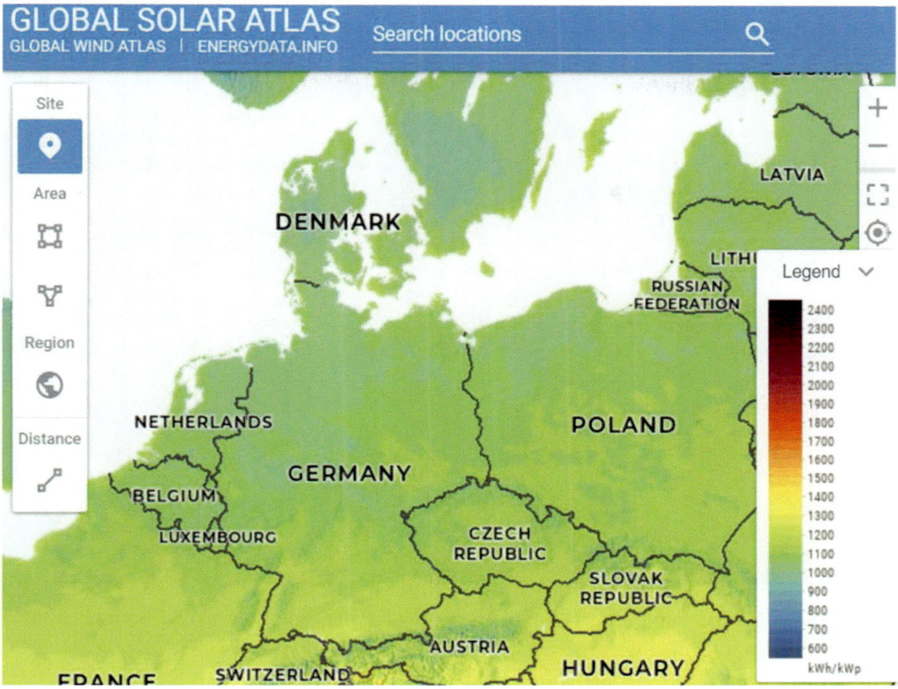

Figure 5.1 Global solar atlas – solar potential of Poland and Germany. Source: https://globalsolaratlas.info/map?c=52.13842,13.857424,5

Firstly, when looking at Poland and Germany displayed on the map below, incorporating global solar atlas data, the two countries tend to have a similar number of kWh/kWp. That metric, simply put, indicates the amount of energy produced (in kWh) during the determined kilowatt 'peak' period (Zhang, 2017), which in this case is in the form of a calendar year. When looking at the values, which are color-coded on the map, both countries have darker green in the northern parts and lighter green mixed with yellow in the south. None of the countries seems to have an advantage over the other in terms of the available solar radiation.

In terms of wind speed, as shown in the visual below, even though it is not enormous, there is a more noticeable variance in colours for the two countries than when looking at the solar atlas. While Poland tends to have evenly distributed wind speed across the land, for Germany, the case is slightly more diverse. On the southern part of that country, the wind speed is more within the lower half of the available range, but the further north one looks, the greater the wind speed. When comparing the two countries, it can be seen that for Poland, the wind farms' potential, when looking solely at

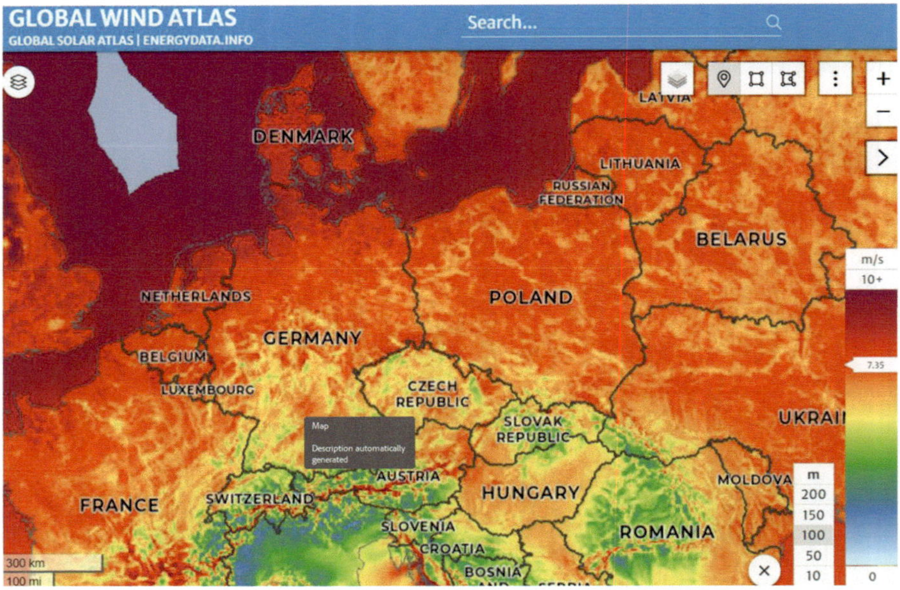

Figure 5.2 Global wind atlas – wind potential of Poland and Germany. Source: https://globalwindatlas.info/

wind availability, is more or less uniform no matter the placement, while for Germany, the northern half of the country seems like a better solution.

5.3.2 Existing energy communities

The European Commission does not specifically define legal and organizational frameworks in which community energy must be managed. Therefore, it is up to each member state to define national regulations that have a direct impact on further concept implementation.

According to the European Directive RED II, there are two specific legal entities recommended for making it possible for citizens to take an active part in the energy system, and those are a renewable self-consumer and a renewable energy community (Appendix 3).

Self-consumer is defined as an individual entity that may generate, store, and sell self-created electricity as long as its main purpose is its own consumption. The community has been characterized by three elements: Open, voluntary, and autonomous participation of members near the renewable energy source; Being either a "natural person," small and medium enterprise, or local authority having "environmental, economic, or social community benefits" priorities over financial profits (European Commission, 2018).

Germany can be perceived as an example of a country where there are no strict regulations for the way energy communities must operate. The concept is often referred to as energy cooperatives or social energy, but none of the term is legally constrained or normalized. On the contrary, Polish legislation on prosumerism and ECs is intensively predefined. In addition to RED II, there is also a third form of participation distinguished, viz, energy cluster. One of the key elements that differentiates it from an EC is the lack of its legal entity. In this form, clusters are functioning as partnerships, in which specific structures and rules are agreed upon on an individual basis in the form of an official agreement between the members. Another characteristic is the larger geographic coverage of clusters, which can be up to five municipalities, compared to three in the case of communities. Greater flexibility of clusters is also provided thanks to the capability of producing electricity for non-members (Kupiec, 2020).

Germany: As mentioned before, Germany places itself as a leading European country in terms of the number of active energy communities. According to information available, this type of cooperative began back in 2006 with eight communities being registered, and up until 2020, there were 888 additional ones established, adding up to 896 energy communities in total. As displayed on the graph below, there was a significant growth in the numbers between 2006 and 2013, as 80% of all the communities we have today were established during that time. The sudden decline after that period was caused by a significant decrease in solar farms' installations and a high increase in wind parks being built, where individuals were no longer favored structures. At that time, the focus shifted towards limited partnerships with limited companies as the preferred option (Wettengel, 2018).

The actual structure of those ECs is characterized by a few factors. The total number of people registered as community members equals 200,000, thus the average EC size is around 230 members. Nevertheless, at the time of a study conducted by DGRV with established ECs, most communities were estimated to have 100–200 members, while one-sixth reported over 500 people involved.

When looking into the inner structure, as displayed above, 95% of ECs in Germany are owned by individuals, with only 1% being associated with public institutions and municipalities. What is more, 80% of the total energy produced by ECs in that country comes from solar panels, indicating its popularity amongst privately-owned communities (DGRV, 2021).

In terms of their distribution across the land, as there are so many privately owned ones, there is not one database containing all of the ECs on

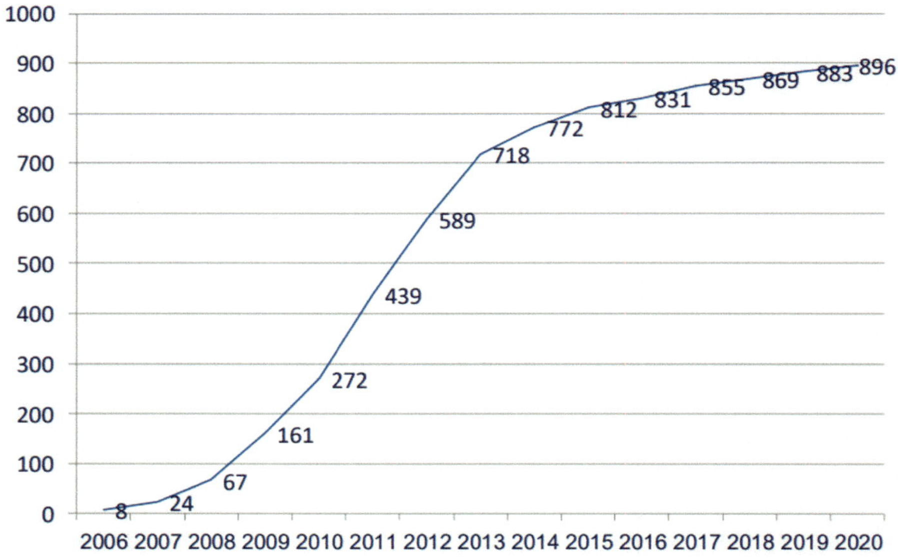

Figure 5.3 Number of energy cooperatives in Germany. Source: https://www.dgrv.de/

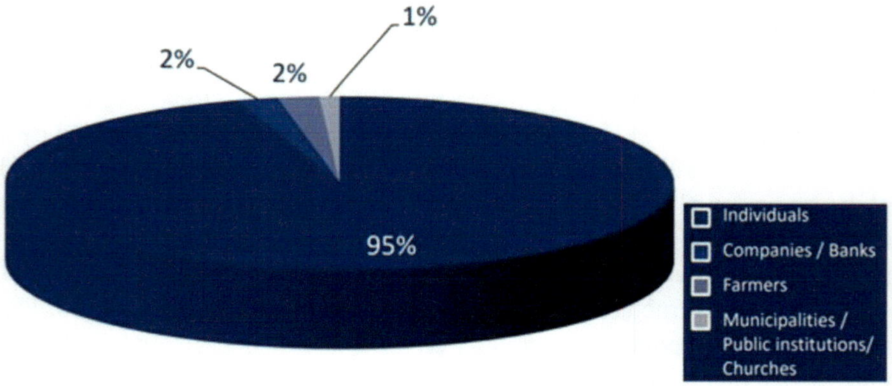

Figure 5.4 Membership structure of energy cooperatives in Germany. Source: https://www.dgrv.de/

one map. Nevertheless, the researchers looked through various websites that comprise information on parts of the overall EC network, and which results can be seen on the three maps below.

On all the maps, there is a visible tendency for more ECs to operate within the western part of the country. Nevertheless, apart from the far north-east area, such communities are present all over Germany. Taking a step back to get a bigger picture of the current energy sector in Germany, it needs to

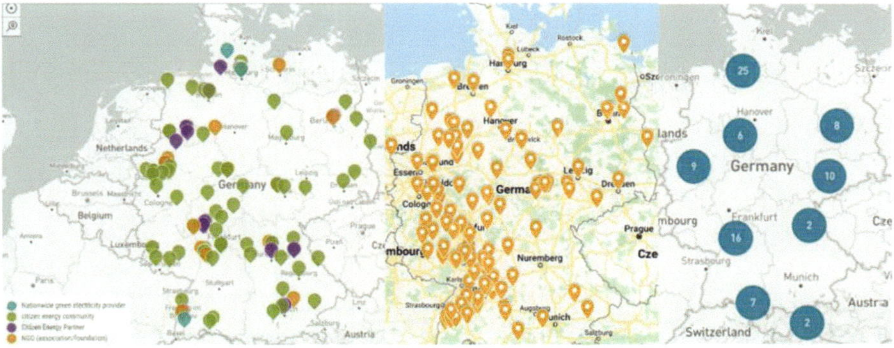

Figure 5.5 Energy Cooperatives in Germany – maps; sources: https://www.buendnis-buergerenergie.de/karte; https://buergerwerke.de/strom-beziehen/die-buergerwerke/die-genossenschaften/; https://www.rescoop.eu/network/map/DE

be mentioned that the share of renewable energy sources is at a rather high level and is only foreseen to increase. In 2021, 46% of the country's energy demand was supplied with RE, with 23% coming from wind and 9,9% from solar (Strom report, 2022). The country plans to increase renewables' share in the energy mix, both in terms of energy consumed and produced (International Energy Agency, 2020). By incorporating guidance included in EU-wide directives, such as RED II, the overall support for EC's development through simplified bureaucratic procedures will be enabled, which is then foreseen to further boost the growth of that sector (Broska et al., 2022).

Combining the above-mentioned information, viz regarding ECs in Germany with the country's energy mix, those 896 communities contribute noticeably to this 41% total renewable energy generation, as they are responsible for 3,5% of that value (DGRV, 2021).

Poland: When looking at the data made available by the International Energy Agency (IEA), the share of coal within the Polish energy power generation sector decreases. However, it is still at a very high level, and the gap between this energy source and any other type is immense. What can additionally be seen on the graph below is that the share of renewables has increased noticeably since 2008, as in 2020 it reached approximately 17% (Ministry of Climate and Environment, 2021). In the official plan for the Polish energy sector, by 2030 RE share is expected to increase to 23%, with noticeable commitment coming from local and civic power generation, which includes ECs (Ministry of Climate and Environment, 2021).

Going into more specific statistics, as shown on the graphic below, 70–80% of the just described RE generated in the country during that year

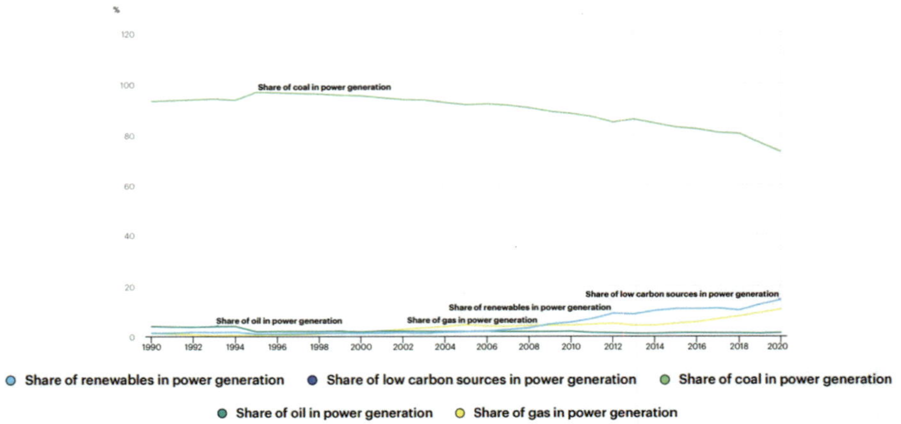

Figure 5.6 Share of renewables, low-carbon sources and fossil fuels in power generation, Poland 1990-2020. Source: IEA

was generated by multiple wind farms across the country. What can be noticed when looking at the wind part of the chart is the substantial increase, up until 2016, in production after which the growth has been stunted (International Energy Agency, 2020). The reason for that finds its explanation in the country's politics, as in that year, President Duda signed a regulation prohibiting the construction of wind turbines within a distance smaller than 2 km from other buildings and natural environments. That bill made it illegal to build a wind turbine across 99% of Poland's total land area, and thus the overall development of that energy sector declined considerably (International Trade Administration, 2021).

As mentioned before, Poland is not a leader when it comes to the number of active and registered ECs. On the contrary, it is situated on the opposite side of the scale, with only two being officially registered so far. The process of establishing new businesses, including ECs, varies significantly from country to country. In Poland, it consists of five main steps, which incorporate approvals from multiple official organizations, as well as three requirements to be fulfilled prior to initiating the official process if one is set to be successful (Krajowy Ośrodek Wsparcia Rolnictwa, 2021).

Starting with the requirements, the EC applying for inclusion in the official list of energy cooperatives needs to:

1. operate in a rural or urban-rural area, but no more than three communes adjacent to each other;
2. have the total number of community members not exceeding 1000;

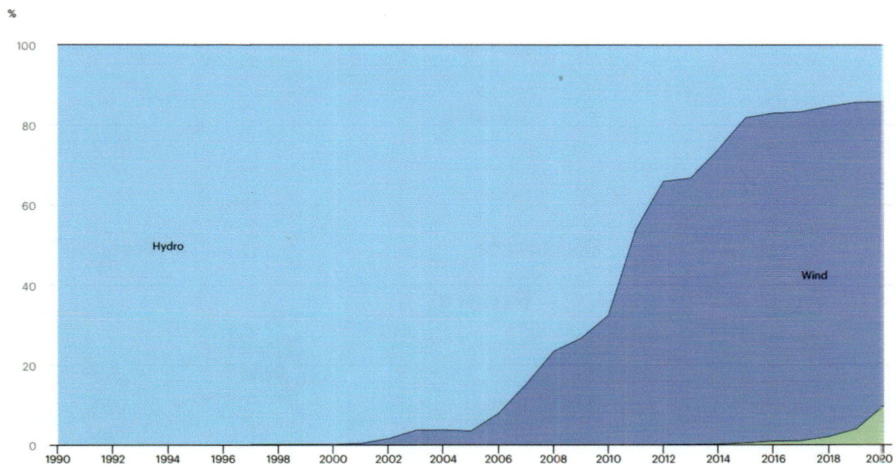

Figure 5.7 Renewable electricity generation by source (non-combustible), Poland 1990–2020. Source: IEA

3. cover not less than 70% of the energy needs of the community and all its members during the year, with the maximum installed capacity of 10 MW when talking about electricity, or have the annual maximum capacity of all installations equal to 30 MW for heat and 40 mln m^3 for biogas.

Once the above steps are fulfilled, the cooperative is entitled to apply to get registered as an operating energy community. That process is initiated by signing the document concerning the statute of the energy cooperative. Once that goes through, the community requests to be enrolled in the Polish National Court Register (KRS), which then leads to the registry in the National Agricultural Support Centre (KOWR). As soon as that part of the process is done, the EC is official. The last part of the process is getting an agreement with the Distribution Network Operator (OSD), which will contain information about the members of the EC, planned measurement systems, and the method of transferring measurement data. With 11 energy cooperatives currently listed in KRS in Poland, from which, as already mentioned, only two have completed the entire process and can be seen within the KOWR registry. The remaining nine are either under liquidation or are not yet accepted by KOWR, due to unknown reasons.

To look for potential patterns for those communities, the researchers decided to create a map on which all those communities are marked. On the map below the blue dots with contact plugs indicate placement for the nine

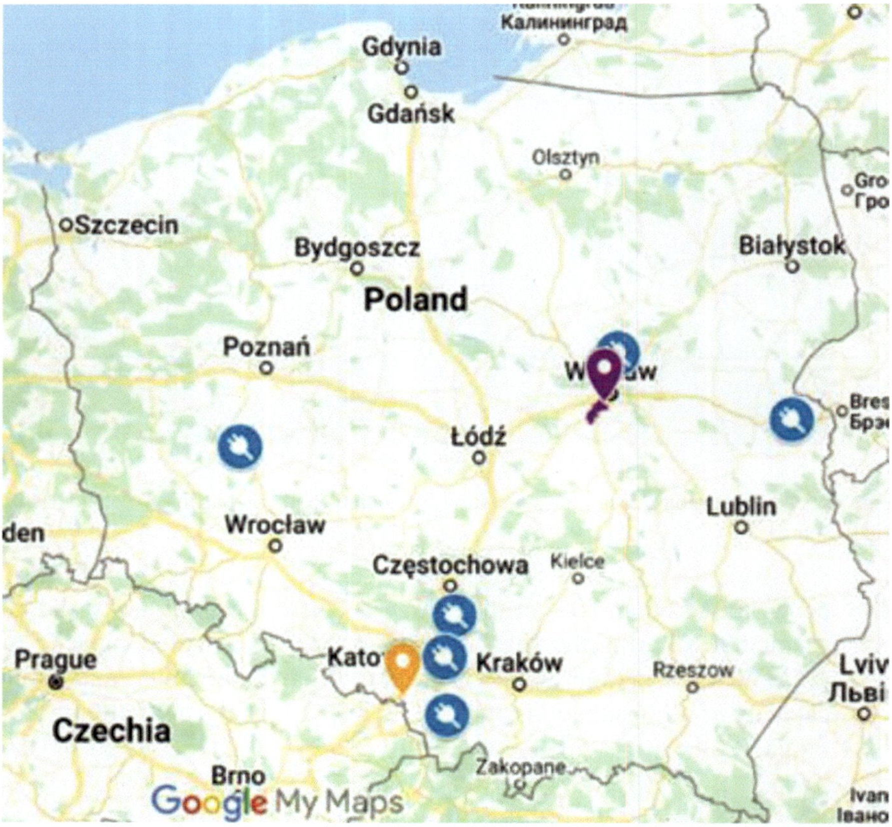

Figure 5.8 Energy Communities in Poland – map. Source: Own production

ECs that are on the KRS list but are not yet approved by KOWR, while dark purple and orange marks display the two official ECs.

To show more in detail what the actual area of the two ECs in Poland is, both of them were zoomed in on the map and are shown below.

The purple region is Eisall EC and consists of three municipalities, viz Nadarzyn, Raszyn, and Michalowice. The orange region refers to 'Nasza Energia' EC and covers the Mszana municipality.

Nevertheless, in the official plan for Polish energy transition, there is a target set to have 300 entities of that sort established by 2030 (Ministry of Climate and Environment, 2021). What was mentioned by the EEC representative was that interest does increase amongst individuals as they are willing to become more independent and in charge of their constantly rising electricity bills coming from the common and publicly available energy grid.

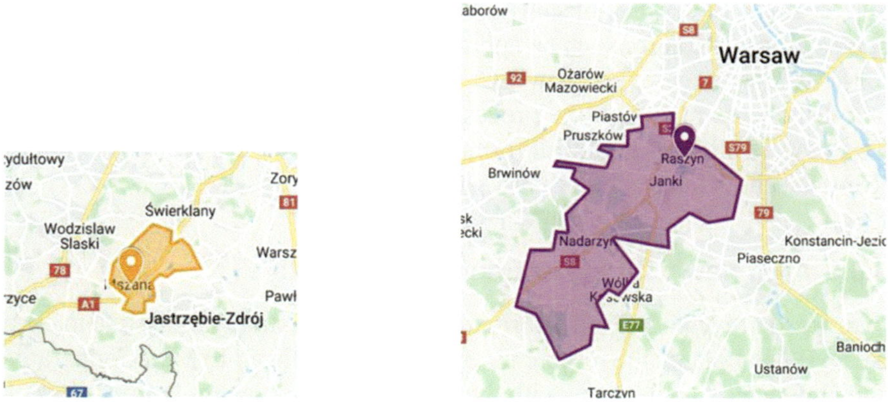

Figure 5.9 Eisall Energy Community and Energy Community 'Nasza Energia' on a map. Source: Own production

One of the initiatives that is being mentioned in terms of supporting the foreseen development of ECs in Poland is a project described as "Rural Development through Renewable Energy Sources in Poland," abbreviated to RENALDO. It is well-understood that investment in innovation and local economies is vital for succeeding with the foreseen energy transition plan, hence why with the help of the European Commission and the German Federal Ministry for the Environment. Its core objective is to advance the expansion of collective prosumer initiatives across rural Poland through the execution of pilot energy cooperative projects across six selected municipalities within Podlaskie and Kuyavian-Pomeranian provinces. The idea behind this initiative is that once the necessary knowledge and details are gathered from those sites, it will become easier to adjust the current legislation and regulatory settings in the country and boost EC's share in the energy transition. The three steps of the project include coming up with an official handbook on how to effectively set up and operate an EC based on two already existing communities, then supporting and development of four additional cooperatives, and lastly analysis of acquired knowledge during the process for future good practices through a common session with the stakeholders (Europäische Klimaschutzinitiative, 2021).

As was mentioned by the representative of Eisall Energy Cooperative (EEC), at the time when EEC wanted to officially become registered as an EC, there were no past experiences with how to do that, as no organization was willing to take the first step and become the lighthouse project earlier. As direct proof of that being the case, having this EC operating within the market for a little bit over 1.5 years now, only last month it managed to get an agreement with OSD in place (Perkowski, 2022). Thus, RENALDO is perceived as a solution for such an approach, where those six municipalities will serve as display cases for how ECs operate and what the potential benefits and disbenefits related to the process are for all stakeholders, so that future ECs have some background information and lessons learned from the past (Europäische Klimaschutzinitiative, 2021).

5.3.3 Interview with Eisall Energy Cooperative

In order to obtain first-hand information on what it actually looks like to operate an EC in Poland, the researchers reached out to Eisall Energy Cooperative and conducted an interview with the Business Development Partner.

What was mentioned the most about the process during that meeting was that it is in fact, very time-consuming and that even though all the already described legal requirements can be difficult to meet, the most constraining ones include having the maximum production capacity lower than 10 MW and addressing over 70% of energy needs with energy coming from EC's grid. What was also emphasized is that not having any other organizations in place prior to initializing the process made it especially difficult for ECC to come through, as there were no templates or tested frameworks to follow, which only postponed the case from progressing. What does not help here is also the fact that there is little to no awareness regarding the EC concept, amongst individuals, but also energy-related businesses and public entities. People tend to be willing to invest if more information about the concept is available at hand.

When asked about whether countries in which ECs tend to be successful, such as Germany, were used as example cases to try and replicate best practices from the industry, the representative emphasised that, as the countries differ a lot, it would not make sense to create such a comparison. With that being the case, a set of rules tailored to the Polish situation was established and is now expected to be followed and fulfilled.

The direct benefits that are closely connected with ECs are both financial and comfort. Primarily, economic as connected to the grid coming from a community tends to get cheaper for the end-user. Although it may seem like having the energy produced at the same expense, there is no additional cost related to its distribution as well and it can be sold back to the grid if more is produced than consumed. In terms of comfort, the matter concerns having a reliable, decentralized, secure, and stable energy source. Those aspects appear to be of high relevance for both private users and the country in general.

5.4 Discussion

The theoretical background, information obtained through the interview, and additional market research emphasized several similarities and differences. The following section of the report will focus on the analysis and comparison of those to establish common findings categorized into politics and economy, technology, and community.

5.4.1 Politics and economy

In the great majority of literature and research material, governmental involvement, its policies, political will, lobbying, as well as local support and governance played an important role in the energy grid mix composites. The RE development in Europe has been strongly correlated with the political support of the members of the cabinet, as well as local government participation. In the case of Polish industry, the focus can be identified as diss-prioritised, if not marginal.

Centrally governed regulations made the country prioritise traditional sources of energy, focused mainly on coal, as well as constraining decentralized RE production investments. The EC concept has been identified as an interesting opportunity, mainly for the rural part of the country, to serve as a means of limited support to the farmland owners to improve their energy independence and financial efficiency. All that, compounded with limited political will and resulted in constrained opportunities for communities to drive the RE transition countrywide on their own.

Some of the supporting projects recognized in the market include RENALDO, which indicates that there are some resources allocated to elevate the concept's implementation in rural areas. Aside from that initiative, there is already increased awareness amongst citizens regarding RES as well as general interest in improving energy efficiency. Nevertheless, limited support via policies, low incentives, recognition, or lack of communication platforms through which knowledge as well as lessons learned could be exchanged, negatively affect the concept's development. These structural and symbolic resources must be developed if further concept implementation is to be achieved.

Until now, it could be noticed that positive narratives have been driven by external entities and local entrepreneurships, while the national government was the responsible party for the core decision-making process. During the research, local administration has been mentioned multiple times as a body that has a growing interest in engagement and yet relatively limited actual power. The local government's involvement has been recognized as potentially beneficial for citizen participation and fulfillment of the current structural and symbolic gap in the resources.

One of the key elements missing in the Polish market is the emphasis on education regarding the importance of ECs and their long-term impact. Citizens' acceptance and support have a direct influence on the overall success of the energy transition. What is more, spreading awareness on the topic of ECs minimized the barrier established before by topic alienation. One of the issues mentioned during the interview conducted was that people are simply not aware that something like an EC is an option. Thus, some of the priority topics, that are to be acknowledged include positive leverage on the local market, increased environmental protection benefits, local community's infrastructure, and status development, as well as long-term economic and energy security.

When looking at information available on Germany and Poland in terms of how to establish an EC, the major difference refers to the actual set of rules and guidelines provided by governments. While for Germany the researchers did not manage to find an official list with steps and requirements to be followed, and the number of registered ECs in that country increases every year, for Poland the process is detailed in an official document. As mentioned by the representative of one of the two existing ECs in Poland, those rules tend to be problematic to fulfil by potential prosumers and might have a direct impact on the slow development of such communities in the country.

5.4.2 Technologies

For both countries, wind and solar can be considered as highly employed sources of RE within their overall energy sectors. When comparing Poland with Germany on solar and wind power potential, solely when looking at the availability of these resources, it can be stated that the countries are on a similar level. Even though Germany displays greater opportunity for wind-generated power within its northern regions, these potential supply variations do not seem to be responsible for the differences visible in terms of RE generation between the two countries. From a purely natural perspective, viz availability of natural resources such as wind and solar, it can be stated that both nations tend to have similar potential for RE farms.

With that being the case, when looking into the actual numbers on ECs, Germany is significantly more developed and mature, as it has nearly 450 times more registered communities currently. Those 896 ECs in Germany contribute to as much as 3.5% of the overall renewable energy produced in the country, while the two communities in Poland cannot be considered on a similar scale. It is evident that technological maturity and access to it play a significant role in countries' energy transition. Hence why many more ECs can be seen in Germany, where RE accounts for over 40% of the energy mix. Nevertheless, even though in Poland the numbers are smaller, the perception of such solutions is positive.

When it comes to potential benefits associated with EC's development, both in literature and conducted research, there were two technological aspects that were frequently mentioned. Firstly, having a decentralized energy grid provides both individual consumers and the overall society with a feeling of security with regards to the energy supply and the overall grid balance. The other element concerns regional innovation and the fact that investing in ECs contributes to everyday life modernization, increased morale, and better infrastructure.

Polish legislations regarding ECs target solely rural areas in the country, where solar energy tends to be the easiest to implement by local authorities for communities on a larger scale. In Germany, on the other hand, even though the energy produced comes mostly from solar, 95% of all ECs are owned by individuals and only 1% is represented by the public. That shows that the Western neighbor has much more freedom and flexibility in terms of where and by whom an EC can be established. Nevertheless, in both cases, EC development is highly dependent on the situation in which the potential community is, including e.g., grid connection and accessibility, which then has a major impact on the way the registration process progresses.

What was also mentioned by Eisall's representative during the interview was the fact that at the moment, EC investment costs in RE technologies are high, and people tend not to even consider making such an investment themselves unless financial incentives from the government appear. Thus, narrowing down EC's scope to only rural regions in Poland may be a reason for not being able to develop many more communities over time, simply due to a lack of financial resources and sufficient incentives. Limiting EC only to rural communities also has an additional negative impact on the RES development in urban societies, where potential market demand in cities is not facilitated.

5.4.3 Community

Finally, the social aspect of the EC implementation has been considered as the end indicator if the concept will be adopted or disregarded.

The research identified the EC concept implementation to be at the starting phase of development in Poland. The perception of the established and soon-to-be-established communities is positive, and those are perceived as "first movers." This indicates that there is a strong will and expectation for continuous efforts in the future, based on the lessons learned, as well as a continuously increasing scale of the projects.

There is still a significant gap between the Polish and German levels of implementation, but the shorter social innovation culture, as well as communal cooperation traditions in Poland, need time to be fully incorporated. However, based on the conducted research, it cannot be argued that the social energy sector in Poland will look the same as Germany. The eastern neighbor is currently driven by different motivations as well and the first movers are coming from different, than what it was in Germany, environments. In the developing market, the main progress channels are entrepreneurs, individuals, and local governments, rather than social groups and local communities. The primary driver motivating the investment is more financial than social capital development and climate conservation, as that is less tangible.

One of the observations that could be driven out of the research is that the benefits that can be gained based on the EC implementations are not disregarded because of a lack of interest from the people, but rather due to a lack of awareness and insufficient knowledge sharing structures. Polish citizens suffer a lack of resources that could help them understand the long-term development impact on the local, as well as national scale, in aspect of energy security and flexibility, labor market, and natural environment preservation. Due to the immaturity of the market, there is limited know-how in the market that could be shared and transferred, which makes it difficult to break the initial barriers. The problem is even compounded by the complexity and unique characteristics of each EC project, which depend on many factors such as demand and supply characteristics of a community.

It has been recognized that there is growing potential in the communities to take an active part in the development of the infrastructure, even if this requires additional financial expenses. This positive objective suffers from a lack of government-driven financial incentives that could boost the engagement and alleviate the high investment costs for the interested parties. Knowledge about potential economic support is especially poorly represented in the main addressee group, viz rural communities, where these kinds of benefits could make the biggest impact due to lower development and financial capabilities of residents.

Finally, it is expected that the involvement of citizens in the energy management system can have a positive effect on the general acceptance of RES development in the country as well as decreased impact of "Not-In-My-Backyard" phenomenon in the citizens' perspectives. By connecting communities to the renewable industry, a direct cost and benefit connection is established, which decreases alienation of the topic for the people as well as creates direct profits and ownership over the RESs. Here, solar energy seems to be the more feasible technology that does not require significant landscape changes, as well as interfere in smaller scale with the additional land and environmental requirements of the citizens.

5.5 Suggestion for Future Research

The research pointed out photovoltaics as the more favorable technology for individual and communal investment, mainly due to the lower investment costs and minimal landscape impact for residents. The perspective, however, does not take into account the perspective brought up in one of the research papers that wind in Poland is assessed to be twice as efficient in energy production capabilities. This characteristic should be deeply considered to prioritize the development of more efficient technologies for the specific geo-locations and communities.

One of the articles reviewed during the primary research proposed a methodology for the most feasible location assessment for the RES development, with Poland as one of the examples. It is recommended to follow the proposed methodology with additional parameters added regarding the location of the grid cable infrastructure, local government location support programs present (or expected to be included in the political program in the future), and proximity to the administrations that have some sort of experience with citizen energy project implementation already.

Limited research focus has been placed on the local government's participation in the RE transition in Poland, thus the topic leaves an opportunity to make a difference in the general topic understanding. It is recommended to continue the EC concept research with a greater focus on the local administration as those are the main stakeholders and organs with higher decision-making power.

Poland is considered a developed country, and yet as a post-communist republic, its socio-economic as well as political situation differs significantly from the rest of the Western member states of the European Union. The cultural and historical contexts have a recognizable impact on the adaptation effect of citizen citizen-driven energy concept; therefore, the socialist past of the country should be studied with additional emphasis in the future.

5.6 Conclusion

The above research was conducted with the core purpose to define the potential of ECs in Poland. The comparison with Germany was made to investigate different aspects that may affect the development of such communities in Poland. The information and data gathered confirmed that Polish western neighbor, even though it displays alike natural conditions and RE potential, is more mature both in terms of RE share in its energy mix and technologies available for individuals. Therefore, it is comprehensible that the development of ECs in Poland occurs at a slower pace.

Nevertheless, based on both primary and secondary data sources, it can be concluded that the concept of ECs in the country is likely to succeed in the future if some parts of the current system are adjusted. The one that was emphasized the most across all the sources concerns the political structure of the process, as it is centrally governed and limits the development of ECs to rural areas and individual municipalities. As the county has set an ambitious goal of having 300 ECs registered by 2030, looking into educating society and spreading awareness regarding the concept and benefits associated with it, as well as potential adjustments to current rules and financial incentives, is considered a necessity.

References

DGRV. (2021). Energy Cooperatives in Germany. Retrieved from State of the Sector 2021 Report: https://www.dgrv.de/wp-content/uploads/2021/06/20210623_ENG_DGRV_Umfrage_Energiegenossenschaften_2021.pdf

Europäische Klimaschutzinitiative. (2021, 04). EUKI Projetcs. Retrieved from RENALDO – Rural Development through Renewable Energy Sources: https://www.euki.de/en/euki-projects/srsp/

European Comission. (2018). EU Science Hub. Retrieved from Renewable Energy – Recast to 2030 (RED II): https://joint-research-centre.ec.europa.eu/welcome-jec-website/reference-regulatory-framework/renewable-energy-recast-2030-red-ii_en

European Comission. (2018, 12 11). EUR-Lex. Retrieved from Directive (EU) 2018/2001 of the European Parliament and of the Council of 11 December 2018 on the promotion of the use of energy from renewable sources: https://eur-lex.europa.eu/legal-content/EN/TXT/?uri=uriserv:OJ.L_.2018.328.01.0082.01.ENG&toc=OJ:L:2018:328:TOC

International Energy Agency. (2020). Germany 2020 - Energy Policy Review. IEA Publications. Retrieved from https://iea.blob.core.windows.net/assets/60434f12-7891-4469-b3e4-1e82ff898212/Germany_2020_Energy_Policy_Review.pdf

International Energy Agency. (2020). Key energy statistics, 2020. Retrieved from Poland: https://www.iea.org/countries/poland

International Trade Administration. (2021, 09 03). Country Commercial Guides. Retrieved from Poland - Energy Sector: https://www.trade.gov/country-commercial-guides/poland-energy-sector#:~:text=Renewable%20energy,-In%202008%20all&text=Poland's%20RES%20capacity%20totaled%209%2C475,in%202020%2C%20reaching%20708%20MW

Krajowy Ośrodek Wsparcia Rolnictwa. (2021). Warunki zamieszczenia danych w wykazie spółdzielni energetycznych. Warsaw: KOWR.

Kupiec, B. (2020). Analiza prawno-porównawcza klastrów energii i spółdzielni energetycznych. Przegląd Prawniczy Uniwersytetu Warszawskiego, 64.93.

Ministry of Climate and Environment. (2021). Energy Policy of Poland until 2040 (EPP2040). Retrieved from https://www.gov.pl/attachment/a1e42067-c749-4dbe-82bf-211d46821d9d

Perkowski, D. (2022, 05 27). Energy communities in Poland - Eisall Energy Community. (A. Krolikowska, & M. Panasik, Interviewers)

Strom report. (2022). Electricity mix. Retrieved from German electricity mix and power generation until 2022: https://strom-report.de/strom/#:~:text=Der%20deutsche%20Strommix%20setzte%20sich,Jahren%20zugunsten%20der%20Erneuerbaren%20verschiebt

Wettengel, J. (2018, 10 25). Clean energy transition - Journalism for the energy transition. Retrieved from Citizens' participation in the Energiewende: https://www.cleanenergywire.org/factsheets/citizens-participation-energiewende#:~:text=farmers%2016%20percent.-,-Energy%20cooperatives,democratically%20with%20many%20equal%20partners

Zhang, T. (2017, 08 14). Solar Power World. Retrieved from What's a good value for kWh/kWp? An overview of specific yield: https://www.solarpowerworldonline.com/2017/08/specific-yield-overview/

Błażejowska M., Gostomczyk W. (2018). Conditions for the Crea tion and State of Development of

Cooperatives and Energy Clusters in Poland Compared with Germany (in Polish). Problems of World

Agriculture, 18(2), 20–32; DOI: 10.22630/PRS.2018.18.2.31

Wawrzyniak, K., Walkowiak, S., & Cetnarski, R. (2021). Elastyczność w sieci OSD jako kluczowy komponent transformacji energetycznej. Energetyka Rozproszona, (5-6), 75–90. https://doi.org/10.7494/er.2021.5-6.75

Jasiński, J., Kozakiewicz, M., & Sołtysik, M. (2021a). Determinants of energy cooperatives' development in rural areas – evidence from poland. Energies, 14(2) doi:10.3390/en14020319

Capellán-Pérez, I., Johanisova, N., Young, J., & Kunze, C. (2020). Is community energy really non-existent in post-socialist europe? examining recent trends in 16 countries. Energy Research and Social Science, 61 doi:10.1016/j.erss.2019.101348

Bomberg, E., & McEwen, N. (2012). Mobilizing community energy. Energy Policy, 51, 435-444. doi:10.1016/j.enpol.2012.08.045

Szewrański, S., Bochenkiewicz, M., Kachniarz, M., Kazak, J. K., Sylla, M., Świąder, M., & Tokarczyk-Dorociak, K. (2019). Location support system for energy clusters management at regional level. Paper presented at the IOP Conference Series: Earth and Environmental Science,, 354(1) doi:10.1088/1755-1315/354/1/012021 Retrieved from www.scopus.com

Jasiński, J., Kozakiewicz, M., & Sołtysik, M. (2021b). The effectiveness of energy cooperatives operating on the capacity market. Energies, 14(11) doi:10.3390/en14113226

Furmankiewicz, M., Hewitt, R. J., & Kazak, J. K. (2021). Can rural stakeholders drive the low-carbon transition? analysis of climate-related activities planned in local development strategies in poland. Renewable and Sustainable Energy Reviews, 150 doi:10.1016/j.rser.2021.111419

Lupi, V., Candelise, C., Calull, M. A., Delvaux, S., Valkering, P., Hubert, W., . . . Dunlop, T. (2021). A characterization of European collective action initiatives and their role as enablers of citizens' participation in the energy transition. Energies, 14(24) doi:10.3390/en14248452

Skoczkowski, T., Bielecki, S., & Baran, Ł. (2016). Renewable energy sources – problems and perspectives of development in poland. [Odnawialne źródła energii – Problemy i perspektywy rozwoju w Polsce] Przeglad Elektrotechniczny, 92(3), 190–195. doi:10.15199/48.2016.03.44

Juszczyk, O., Juszczyk, J., Juszczyk, S., & Takala, J. (2022). Barriers for renewable energy technologies diffusion: Empirical evidence from finland and poland. Energies, 15(2) doi:10.3390/en15020527

Drożdż, W., Mróz-Malik, O., & Kopiczko, M. (2021). The future of the polish energy mix in the context of social expectations. Energies, 14(17) doi:10.3390/en14175341

Markuszewska, I. (2021). 'Old trees cannot be replanted': When energy investment meets farmers' resistance. Journal of Settlements and Spatial Planning, 2021(Special Issue 8), 5-13. doi:10.24193/JSSPSI.2021.8.02

Krawczak, E. (2019). Studies on PV power plant designing to fulfil the energy demand of small community in poland. Paper presented at the E3S Web of Conferences, 116 doi:10.1051/e3sconf/201911600040 Retrieved from www.scopus.com

Dragan, D. (2020). Legal barriers to the development of energy clusters in poland. European Energy and Environmental Law Review, 29(1), 14–20. Retrieved from www.scopus.com

Hielscher, S., Wittmayer, J. M., & Dańkowska, A. (2022). Social movements in energy transitions: The politics of fossil fuel energy pathways in the united kingdom, the netherlands and poland. Extractive Industries and Society, doi:10.1016/j.exis.2022.101073

Cichowska, J. B., Mreła, A., & Sokolov, O. (2021). The importance of the renewable energy sector in the opinion of rural and urban youth. [Znaczenie sektora energetyki odnawialnej w opinii młodzieży wiejskiej i miejskiej] Polityka Energetyczna, 24(2), 157-181. doi:10.33223/epj/137962

Appendix 3 – Chapter 5

RED II review

Selected sections of the Renewable Energy Directive 2018/2001/EU (RED II), relevant to the energy communities topic.

https://eur-lex.europa.eu/legal-content/EN/TXT/?uri=uriserv:OJ.L_.2018.328.01.0082.01.ENG&toc=OJ:L:2018:328:TOC

Preamble from points 63 onwards:

63 – Investment in RE supports local development including labor and technological advancement, as well as stimulates the development of SMEs, prosumers, and energy communities' activity

64 – Remote locations can benefit from ET significantly due to the limited access to grid and fossil fuel-based energy production, those are expected to have favorable access to RE sources, especially biomass that could be utilized for bio-liquid and bio-gas fuels.

65 – Decentralized RE technology development should be managed in a financially available way for the local investment, which is expected to bring many benefits such as utilization of renewable resources, energy security, labor market development, shortened energy transportation and loss, increase in social capital.

66 – Push towards re-definition of prosumers and energy communities to decrease unnecessary burdens

70 – Citizen participation in investment towards RE developments through energy communities increases acceptance rates of RE technology and additionally increases the speed of development

71 – Energy communities and prosumers should be able to trade and manage their energy within the energy system, without discrimination and favoring of any parties

Art 2 definitions – direct citation

14 – "Renewables self-consumer" means a final customer operating within its premises located within confined boundaries or, where permitted by a member state, within other premises, who generate renewable electricity for its

own consumption, and who may store or sell self-generated renewable electricity, provided that, for a non-household renewables self-consumer, those activities do not constitute its primary commercial or professional activity;

16 – "Renewable energy community" means a legal entity:

(a) which, in accordance with the applicable national law, is based on open and voluntary participation, is autonomous, and is effectively controlled by shareholders or members that are located in the proximity of the renewable energy projects that are owned and developed by that legal entity;

(b) the shareholders or members of which are natural persons, SMEs, or local authorities, including municipalities;

(c) the primary purpose of which is to provide environmental, economic or social community benefits for its shareholders or members or for the local areas where it operates, rather than financial profits.

Art 21 – Renewables self-customers

3c) prosumers can be obliged to pay non-discriminating fees if the total energy power installed is above 30 kW

Art 22 – Renewable energy communities

4 – Countries are responsible for making sure that communities do not have unnecessary difficulties with opening, as well as operating, RE community

6 – It is possible that energy communities also operate transborder

6

Concluding Remarks on Energy Inequalities

George Xydis[1,2]

[1]Department of Mechanical Engineering, University of the Peloponnese, Greece
[2]Department of Business Development and Technology, Aarhus University, Denmark

6.1 Conclusion

The book *Energy Inequalities: Crises, Pricing, and Policies* underscores the urgent need to address the systemic disparities shaping today's energy systems. From the unequal burdens of energy crises on vulnerable households to the risks of exclusion in the transition to electric vehicles and the untapped potential of energy communities, this book highlights how policy, market structures, and infrastructure either perpetuate or mitigate inequality. The case studies reveal that while behavioral adaptations and grassroots initiatives offer promising pathways, structural barriers – regulatory inertia, financial constraints, and centralized governance – often hinder equitable progress.

Moving forward, achieving a just energy transition demands more than technical solutions; it requires policies that prioritize inclusivity, empower communities, and redistribute the costs and benefits of energy systems fairly. The insights and exercises presented in this book serve as a foundation for policymakers, researchers, and advocates to confront these challenges head-on. By centering equity in energy decision-making, we can build resilient systems that serve not just the privileged few. The time for action is now.

DOI: 10.1201/9788743808923-6

Appendix A

Exercises on Energy Inequalities and Energy Policies for University Students

George Xydis

[1]Department of Mechanical Engineering, University of the Peloponnese, Greece
[2]Department of Business Development and Technology, Aarhus University, Denmark

Introduction to exercises

Practice makes progress

This chapter is your hands-on toolkit for mastering energy inequalities. Unlike traditional textbook sections, every page here is designed for active learning – through problem-solving, application tasks, and critical thinking challenges.

How to use this chapter:

- Skill-building journey: Exercises progress from foundational to advanced – complete them in order for maximum benefit.
- Self-assessment: Check your understanding as you go using the answer key (Solution following).
- For instructors: These can be assigned as homework, in-class activities, or discussion starters.

Why it works:

- Learn by doing: Apply theories immediately to reinforce concepts.
- Mistakes welcome: Struggle is part of the process. Space is provided for drafts and revisions.

- Real-world ready: Many exercises simulate professional or academic scenarios.

Exercise A.1 Quantifying EV charging cost disparities

Calculate and compare annual electricity costs for different consumer groups under time-of-use (ToU) pricing, evaluating how EV adoption exacerbates energy inequalities.

Given data:

Tariff structures (hypothetical but realistic):

✓ Flat Rate: €0.20/kWh (all hours)

✓ Time-of-Use (ToU) Rate:

  - Peak (4 PM–9 PM): €0.30/kWh
  - Off-Peak (9 PM–8 AM): €0.10/kWh
  - Shoulder (8 AM–4 PM): €0.15/kWh

User profiles:

- Household A:

Income: High
EV charging: Home charger, 100% off-peak (overnight)
Monthly EV consumption: 300 kWh
Baseline household use: 500 kWh (30% peak, 50% off-peak, 20% shoulder)

- Household B:

Income: Low
EV charging: Public fast-charger, 70% peak (after work), 30% shoulder
Monthly EV consumption: 300 kWh
Baseline household use: 500 kWh (50% peak, 30% off-peak, 20% shoulder)

Additional constraints:

- Public fast-chargers add a €0.05/kWh service fee.
- Low-income households lack rooftop solar or battery storage.

Questions:

1. Calculate annual electricity costs for both households under:
   - Flat-rate pricing.
   - TOU pricing.

(Assume no seasonal variations.)

2. Cost increase due to EV integration:
   - Compare the percentage increase in annual bills for Households A and B after adding EV loads.
3. Inequity analysis:
   - Compute the disparity ratio: (Household B's cost)/(Household A's cost) under ToU.
   - How much more does Household B pay annually due to charging behavior constraints?

Exercise A.1 Indicative brief solution

1. Household A (high-income) calculations:
   ToU baseline use:
   Peak: 500 × 30% × 0.30 = €45
   Off-peak: 500 × 50% × 0.10 = €25
   Shoulder: 500 × 20% × 0.15 = €15
   Total baseline: €85/month
   EV charging (off-peak): 300 × 0.10 = €30
   Total monthly (ToU): 85 + 30 = €115

2. Household B (low-income) calculations:
   ToU baseline use:
   Peak: 500 × 50% × 0.30 = €75
   Off-Peak: 500 × 30% × 0.10 = €15
   Shoulder: 500 × 20% × 0.15 = €15
   Total baseline: €105/month

   EV charging:
   Peak: 300 × 70% × (0.30 + 0.05) = €73.50
   Shoulder: 300 × 30 = €73.50
   Shoulder: 300 × 0.3 × (0.15 + 0.05) = €18 (because public charger markup: + €0.05/kWh (on top of ToU rates)
   Total EV cost: €91.50
   Total monthly (ToU): 105 + 91.50 = €196.50

3. Disparity results:
   Annual cost (ToU):
   Household A: 115 × 12 = €1380

Household B: 196.50 × 12 = €2358
Disparity ratio: €2358/€1380 = 1.71 (Household B pays 71% more).

Exercise A.2 Grid Fee Allocation in Renewable Energy Communities (RECs)

An exercise focused on energy community cost-sharing models and grid fee disparities, incorporating real-world pricing structures and equity analysis. Calculate how different grid fee structures impact member bills in a solar+EV energy community and evaluate fairness. The given data are:

- Community setup: 10-household REC (5 high-income with rooftop solar + EVs, 5 low-income with no solar/EVs).
- Total annual energy consumption: 50,000 kWh (shared equally: 5000 kWh/household).
- Solar generation: 30,000 kWh/year (owned by high-income households).

Cost parameters:

- Retail electricity rate: \$0.25/kWh (from grid).
- Solar Feed-in Tariff: \$0.10/kWh (sold to grid).

Grid fees:

- Option 1: Fixed \$100/year per household.
- Option 2: Per-kWh fee: \$0.05/kWh on all grid imports.
- Option 3: Capacity-based fee: \$10/month per kW of grid connection capacity (all households: 5 kW).

EV loads:

- High-income households each charge an EV (2000 kWh/year, all off-peak).

Calculate the annual costs for each household under all 3 fee structures:

- Assume solar is shared equally among REC members (pro-rata).
- Low-income households import 100% of power from the grid; high-income households use solar first.

What do you notice?

Exercise A.2 Indicative Brief Solution

Option 1: Fixed Fee ($100/Household)

High-income household:
Solar used locally: 3000 kWh (of 5000 kWh needed).
Grid imports: 2000 kWh (including EV) × $0.25 = $500.
Revenue from excess solar: (3000 kWh – 2000 kWh) × $0.10 = $100.
Total cost: $500 (grid) + $100 (fee) – $100 (revenue) = $500/year.

Low-income household:
All 5000 kWh from grid × $0.25 = $1,250.
Total cost: $1,250 + $100 = $1350/year.

Option 2: Per-kWh Fee ($0.05/kWh Imported)

High-income household:
Grid imports: 2000 kWh × ($0.25 + $0.05) = $600.
Revenue: $100 (as above).
Total Cost: $600 – $100 = $500/year.

Low-income household:
5000 kWh × ($0.25 + $0.05) = $1500/year.

Option 3: Capacity-based fee ($10/month × 5 kW)

All households pay $60/year (5 kW × $12).
High-income cost: $500 (grid) + $60 – $100 = $460/year.
Low-income cost: $1,250 + $60 = $1,310/year.

Comment: Low-income people pay a lot more compared to high-income people.

Exercise A.3 Dynamic pricing and energy poverty in EV-dominated grids

Model how real-time pricing (RTP) exacerbates energy poverty when low-income households lack access to demand flexibility. The given data are:

Household profiles:

Household W:

- Income: $100,000/year
- Assets: Rooftop solar (5 kW), home battery (10 kWh), EV (V2G-enabled).

- Demand flexibility: Shifts 80% of peak load to off-peak.

Household P:

- Income: $25,000/year
- Assets: No solar/battery, old EV (no V2G), rigid demand.
- Pricing structures:
  - Flat Rate: $0.18/kWh
  - Real-Time Pricing (RTP):

Hour type	Price ($/kWh)
Peak (5–8 PM)	$0.45
Off-peak (12 AM–6 AM)	$0.08
Shoulder	$0.15

Load profiles (Weekday, Summer):

Household W:

Base load: 20 kWh (30% peak, 50% off-peak, 20% shoulder).
EV charging: 15 kWh (100% off-peak).
Solar generation: 25 kWh (70% during peak hours).

Household P:

Base load: 20 kWh (50% peak, 30% off-peak, 20% shoulder).
EV charging: 15 kWh (70% peak, 30% shoulder).
V2G assumptions: Household W sells 10 kWh back to the grid during peak at $0.45/kWh.

Compute electricity costs for both households under the flat rate and RTP.

- Energy burden analysis: Calculate annual energy burden (percentage of income spent on electricity) for both households under each pricing scheme.
- Quantify the "flexibility penalty" (additional cost to Household P due to rigid demand) under RTP.

A proposed "Energy Justice Tariff" offers:

- Low-income households: Peak price capped at $0.25/kWh.

- Funded by a 5% surcharge on V2G revenue.
- Recalculate costs and burdens.

Exercise A.3 Indicative brief solution (and reflection)

Solution (Household P Under RTP):

Base load cost:
Peak: 20 kWh × 50% × \$0.45 = \$4.50
Off-Peak: 20 kWh × 30% × \$0.08 = \$0.48
Shoulder: 20 kWh × 20% × \$0.15 = \$0.60
Total base: \$5.58

EV charging cost:
Peak: 15 kWh × 70% × \$0.45 = \$4.73
Shoulder: 15 kWh × 30% × \$0.15 = \$0.68
Total EV: \$5.41

Total daily cost (RTP): \$5.58 + \$5.41 = \$10.99
Under flat rate: (20 + 15) kWh × \$0.18 = \$6.30
Annual energy burden (RTP): {(\$10.99×365)/\$25,000} ×100%=16% (vs. 9.2% under flat rate)

Critical insights:

- RTP disproportionate impact: Household P's burden rises 74% under RTP, while Household W's burden decreases due to V2G profits.
- V2G as an inequity driver: Household W earns \$4.50/day (\$1,642/year) by selling energy, while Household P pays \$4.69/day more under RTP.
- Policy effectiveness: The capped peak price reduces Household P's burden to 12.1% but requires taxing V2G revenue.

Ethical dilemma: Should V2G participants subsidize low-income households, or does this discourage EV adoption?

Exercise A.4 Rooftop solar adoption and grid cost shifting

Quantify how unequal access to rooftop solar exacerbates energy burdens for non-adopters through grid cost shifts. Utilities recover grid maintenance costs via volumetric electricity rates (\$/kWh). When high-income households install solar and reduce grid purchases, fixed costs are redistributed to remaining users – often low-income households without solar access. The given data are:

Community profile:

- 1000 households:
  - Group A (20%): High-income, owns rooftops, 80% installs 5 kW solar systems.
  - Group B (80%): Low-income, renters, 0% solar access.
- Pre-solar grid cost recovery: $200,000/year via $0.20/kWh rate.
- Average pre-solar use: 5000 kWh/household/year.

Solar performance:

- Each 5 kW system generates 6000 kWh/year (offsetting grid use 1:1).
- Utility still incurs $100/year/household in fixed costs (wires, transformers).

Calculate:

- Calculate pre-solar equity:
  - Show total revenue from the grid ($200,000) and average household payment.
- Post-solar cost shift:
  - Compute new grid purchases after solar adoption.
  - Determine the new kWh rate needed to recover $200,000 + fixed costs ($100 × 1000).
  - Calculate annual costs for:
    - Solar-owning households (paying new rate on residual use).
    - Non-solar households.
- Inequity analysis:
  - Compare % of income spent on electricity for both groups (assume:
    - Group A income: $100,000/household.
    - Group B income: $30,000/household.

Exercise A.4 Indicative brief solution

1. Pre-solar baseline:
   Total grid revenue: 1000 HH × 5000 kWh × \$0.20 = \$1,000,000 (covers \$200k costs + \$800k generation).

2. Post-solar changes:
   - Group A (200 HH):
     - 160 HH install solar → generate 160 × 6,000 kWh = 960,000 kWh.
   - Group B (800 HH): Still uses 5000 kWh each.
   - Total grid purchases post-solar: 800 × 5000 = 4,000,000 kWh.
   - New rate calculation:
     - Fixed costs: \$200,000 + (1000 × \$100) = \$300,000.
     - New rate: \$300,000 ÷ 4,000,000 kWh = \$0.075/kWh (but utilities still need \$800k for generation!)
     - Adjusted rate: \$0.075 + \$0.20 (gen) = \$0.275/kWh.

3. Cost impacts:
   - Group A: Pays \$0.275 × 0 kWh + \$100 fixed = \$100/year (was \$1000).
   - Group B: Pays \$0.275 × 5000 = \$1,375 + \$100 = \$1,475/year (was \$1000).
   - Energy Burden Increase:
     - Group B: \$1,475 ÷ \$30,000 = 4.9% (from 3.3%).
     - Group A: \$100 ÷ \$100,000 = 0.1% (from 1%).

Index

About the Author

George Xydis had the opportunity to work in the wind sector together with developers, utilities, constructors, universities, and research institutes. He used to work as a Wind Projects Development Coordinator at Iberdrola Renewables, as a Wind Project Developer at Vector Hellenic Windfarms S.A., and as a Researcher at the Center for Electric Power and Energy, Department of Electrical Engineering at the Technical University of Denmark. Since February 2017 he has been with the Center for Energy Technologies, Aarhus University as an Associate Professor and since January 2023 a Full Professor. George was one of the co-founders of VerdeCube and worked as an adjunct faculty member at Johns Hopkins University in Energy Policy and Climate program, teaching Wind Energy: Science, Technology and Policy. Since January 2024 he has also been with the University of the Peloponnese, Dept. of Mechanical Engineering.

He used to collaborate, as a freelancer, with institutes, universities, and SMEs. George's research interests include energy planning, wind resource assessment, technoeconomics of renewable energy systems, energy demand and plant factories. He has co-authored more than 140 publications in international journals, more than 50 publications in international conferences, books, and book chapters, and serves as an Associate Editor in Heliyon Energy, Frontiers in Energy Systems & Policy, Journal of Business Analytics, and Energy & Environment. Since May 2024, he has served as the Co-Editor in Chief of Cambridge Prisms: Energy Transitions, a scholarly journal that Cambridge University Press launched. He holds a PhD from the National Technical University of Athens, and a degree in Mechanical Engineering from the Aristotle University of Thessaloniki, Greece.

For Product Safety Concerns and Information please contact our
EU representative GPSR@taylorandfrancis.com Taylor & Francis
Verlag GmbH, Kaufingerstraße 24, 80331 München, Germany